# Trace Theory for
# Automatic Hierarchical Verification of
# Speed-Independent Circuits

## ACM Distinguished Dissertations

**1982**
*Abstraction Mechanism and Language Design*
Paul N. Hilfinger
*Formal Specification of Interactive Graphics Programming Language*
William R. Mallgren
*Algorithmic Program Debugging*
Ehud Y. Shapiro

**1983**
*The Measurement of Visual Motion*
Ellen Catherine Hildreth
*Synthesis of Digital Designs from Recursion Equations*
Steven D. Johnson

**1984**
*Analytic Methods in the Analysis and Design of Number-Theoretic Algorithms*
Eric Bach
*Model-Based Image Matching Using Location*
Henry S. Baird
*A Geometric Investigation of Reach*
James U. Korein

**1985**
*Two Issues in Public-Key Cryptography*
Ben-Zion Chor
*The Connection Machine*
W. Daniel Hillis

**1986**
*All the Right Moves: A VLSI Architecture for Chess*
Carl Ebeling
*The Design and Evaluation of a High Performance Smalltalk System*
David Michael Ungar

**1987**
*Algorithm Animation*
Marc H. Brown
*The Rapid Evaluation of Potential Fields in Particle Systems*
Leslie Greengard

**1988**
*Computational Models of Games*
Anne Condon
*Trace Theory for Automatic Hierarchical Verification of Speed-Independent Circuits*
David L. Dill

# Trace Theory for
# Automatic Hierarchical Verification of
# Speed-Independent Circuits

David L. Dill

The MIT Press
Cambridge, Massachusetts
London, England

This book was printed and bound in the United States of America.

Library of Congress Cataloging-in-Publication Data

Dill, David L.
   Trace theory for automatic hierarchical verification of speed-independent circuits / David L. Dill.
      p. cm.—(ACM distingished dissertations)
   Bibliography: p.
   Includes index.
   ISBN 0-262-04101-4
   1. Switching circuits. 2. Integrated circuits—Very large scale integration. 3. Sequential machine theory. I. Title. II. Series.
   621.39'5—dc20                                      89-12546
                                                          CIP

*To my parents*

# Table of Contents

# Figures

# Series Foreword

The doctoral thesis presented in this volume was written by David Dill and has been honored as Distinguished by the selection committee for the 1988 Distinguished Doctoral Dissertation Award sponsored by the Association for Computing Machinery and The MIT Press.

This distinguished dissertation competition began in 1982 to identify the finest examples of research and scholarship among the doctoral theses submitted each year in computer science and engineering. The competition has become steadily sterner. This year more than 50 nominations were received from top research universities around the world. To be acclaimed in the presence of such competition truly distinguishes Dr. Dill's dissertation.

Advised by Professor Edmund Clarke and motivated by the rapidly increasing complexity of VLSI chip designs, Dr. Dill studied the problem of establishing the correctness of circuits. Traditionally circuit designs have been shown correct by building a prototype and testing it with specific inputs, but as circuits become more complex a smaller proportion of the combinatorily explosive state space can be examined in a realistic amount of time. Furthermore, for the asynchronous, clockfree circuits known as speed-independent circuits, timing adds another dimension that dramatically increases the complexity of testing. Dill applies verification methodology to prove the circuits correct.

At the core of the new theory he developed and implemented is the concept a *trace structure* which represents a circuit's behavior as input/output transitions. Trace structures serve as a formalism for expressing both descriptions and specifications. This double duty not only simplifies the theory but it makes it naturally hierarchical, since specifications at one abstraction level can be descriptions at higher levels. By exploiting such advantages, Dill's demonstration verifier has discovered bugs in published circuit designs.

Lawrence Snyder
Chair, ACM Distinguished Doctoral Dissertation
Award Selection Committee

# Preface

In the continuing quest for faster computation, it is inevitable that parallelism will be heavily exploited. Unfortunately, unexpected interactions among processes running in parallel has been a rich source of subtle errors for a long time. It has become apparent that understanding concurrent systems is vital if we are to have confidence in our computers (and the airplanes and toasters that use them).

This thesis describes an attempt to understand concurrency in a specialized class of systems: speed-independent circuits (sequential circuit designs that do not depend on the magnitudes of the delays in their components). However, we study this narrow class of systems very deeply — we begin by determining the appropriate algebraic properties for formal semantics of circuits, develop a formal semantics, define what it means to implement a specification, and prove that the verification problem is decidable. We then describe an implementation and, as an example, describe how it uncovered a previously unknown bug in a published circuit design. Then we extend the formal model to handle liveness properties, which the first model is unable to do.

Thinking about circuits has turned out to be very productive in complexity theory. Perhaps the same will be true of concurrency, for similar reasons. They allow one to focus on issues of communication and synchronization without distractions such as variable binding, parameters, and so on, that exist in programming languages. At least two new concepts emerged in this study: *conformation* is a binary relation between behaviors that holds when the first can be safely substituted for the second; *receptiveness* is a property of a specification that says, intuitively, whether the specification allows all possible inputs to occur (any specification that does not satisfy this property cannot be implemented, and has other unfortunate properties as well). The decision procedure for this property makes use of some old results from automata theory that, to the author's knowledge, had not previously been applied to linear-time models.

Although this study covers a certain amount of theory, it was intended to be of practical value. We hope that the results herein will ultimately contribute to the intellectual and computational tools that will be necessary for asynchronous design styles to realize their potential.

# Acknowledgements

As always, this work would not have been possible without the help of many other people. First among these is my graduate and thesis advisor, Ed Clarke. Ed adopted me after I was orphaned (so to speak) by the departure from CMU of my previous advisor. I learned what I know about conducting research primarily from him. I first looked into the verification of asynchronous circuits on Ed's suggestion, and my first published results were achieved by analyzing a circuit that he suggested as an example. He encouraged, listened, and occasionally prodded throughout the entire duration of my thesis work.

The other members of my committee, Randy Bryant, Bob Sproull, and Chuck Seitz, were unusually helpful, without exception. Each of them contributed knowledge of the subject area in addition to insights about the thesis itself. They also read drafts quickly and carefully.

I would especially like to thank Bob Sproull, whose enthusiasm for the program of Chapter 5 lifted my morale greatly — not only did he run the program on many of his own designs, he extended it and re-implemented parts of it (his changes are not reported here).

Alain Martin challenged me to verify his distributed mutual-exclusion circuit in 1985 and several times thereafter. I finally did, two years later, with the results reported in Chapter 5. Alain also sent me a revised version of the circuit and clarified the assumptions under which it was to be verified.

A crucial step in the evolution of this work, including the decision to use trace theory, was Sproull and Sutherland's asynchronous circuits seminar, held in the Fall semester of 1985. All of the participants, especially Ivan Sutherland, Charles Molnar, Bob Sproull, and Ed Frank, taught me a great deal about the theory and practice of asynchronous circuit design at an ideal time.

During the seminar, I had several discussions with Charles Molnar; an insight during one of these led to the key idea of explicit failure traces in trace structures.

David Black was another participant in the seminar who, like me, was interested in extending trace theory to capture liveness and fairness properties. David and I worked on various problems which were, in essence, a decision procedure for the receptiveness of complete trace structures. It was in a discussion of this question with David that I came up with the intuition behind the characterization in terms of infinite games that is given in Chapter 7. David also saved me a great deal of time and effort by tutoring me in existing results in trace theory.

Allen Emerson referred me to a discussion of infinite games in Trakhtenbrot and Barzdin's book; this allowed me to prove the decidability result in Chapter 6.

Previous investigators of trace theory for asynchronous circuits, namely Jan van de Snepscheut, Jan Tijman Udding, Jo Ebergen, Huub Schols, Martin Rem, and Tom Verhoof, have all been helpful in explaining their results and viewpoints to me. Martin Rem asked a question which eventually led to the inclusion of the sections on lattice properties of trace structures.

Steven Brookes contributed by explaining CCS and related topics on several different occasions, and by reading some partial drafts of the thesis. Steve also helped me to clarify the idea of "substitution in a context" in Section 4.2.

Pam Gage drew early versions of many of the diagrams in the thesis. She also reminded me to use topic sentences in my paragraphs. At least some of the paragraphs now have topic sentences.

I know also that I have had many discussions with others, too numerous to list, that eventually helped to clarify the presentation of the ideas in this dissertation. A generic "thanks" to you all.

I would like to thank the various support people at CMU, facilities, operations, and administration, for providing me with the resources to complete this work while insulating me from the accompanying hassles. Along these lines, I would especially like to thank Alfred Spector, who loaned me a workstation during the writing and formatting of the thesis, speeding my progress substantially.

Finally, there was some money involved: this research was partially supported by NSF Grant MCS-82-16706 and by the Defense Advanced Research Projects Agency, ARPA Order Number 4976.

Of course, the views and conclusions contained in this document are those of the author and should not be interpreted as representing the official policies, either expressed or implied, of the Defense Advanced Research Projects Agency or the US Government.

# Trace Theory for
# Automatic Hierarchical Verification of
# Speed-Independent Circuits

# Chapter 1

# Introduction

## 1.1. Introduction

VLSI and concurrency are two of the most active areas of computer science research. In VLSI, the technological trend is towards more complex and faster circuits. There is also a continuing challenge to reduce the time and expense of producing correct designs. The challenge in concurrent systems is correctness: because they are inherently nondeterministic, they are much trickier than sequential circuits.

### 1.1.1. Asynchronous Circuits

This dissertation explores a topic at the convergence of VLSI and concurrency: the semantics and automatic verification of asynchronous circuits. Asynchronous circuit design, after many years of neglect by all but a few engineers and researchers, has recently attracted renewed interest within the VLSI community because it offers a way to cope with problems that accompany the increasing complexity of VLSI circuits. Asynchronous circuits are also interesting to study because they are pure concurrent systems. Unlike concurrent programming languages, which present various distractions such as complicated data domains, variable binding constructs, and recursion, an asynchronous circuit consists of a fixed set of modules communicating over a fixed set of unbuffered channels (wires) with no hidden handshaking or other implicit protocols. The complexity of asynchronous circuit operation stems entirely from communication and synchronization.

Timing is a central problem in sequential circuits. The circuit design must ensure that the computation of values has completed before the values are stored. In synchronous circuits, storage elements latch their inputs when a tick of a global clock occurs. Some upper bound on the delays involved in computing values must be estimated, measured, or guessed, then the clock ticks are spaced to allow at least this amount of time.

An *asynchronous circuit* is a sequential digital circuit with no clock. Timing problems in asynchronous circuits must be avoided by carefully coordinating the various delays in the circuit, or by using synchronization signals between circuits. Much of the work in formal models and synthesis of asynchronous circuits was done between the late 1950's to the middle 1960's. Since then, synchronous design methods have been overwhelmingly favored, primarily because of the comparative difficulty of asynchronous circuit design.

The advent of VLSI has stimulated renewed interest in speed-independent design. As circuits become more complex, the cost of communications become increasingly important. Broadcasting a global signal, such as a clock, becomes particularly troublesome. Events which are supposed to be simultaneous can occur at significantly different times because of transmission delays. When the events are clock ticks, this phenomenon is called *clock skew*; it can be a serious problem in the design of large systems. One solution to these problems is to use *speed-independent* components — asynchronous circuits that do not depend on the relative speeds of components to which they are connected.

Speed-independent circuits can also expedite circuit design. Since their correct operation is not sensitive to variations in timing due to process variation, layout, and other aspects of lower-level implementation, they are more likely to work properly when fabricated. They can also reduce design time by simplifying the overall process. For example, a change in the fabrication process or a modification in the layout would not necessitate re-timing the clock.

Finally, VLSI has made it easier to design asynchronous circuits. One of the difficulties in asynchronous design has been the unavailability of appropriate components in mass-produced small- and medium-scale integrated circuits. VLSI design tools have expanded the menu of components that a designer can use — new primitives can be custom-designed, if necessary.

### 1.1.2. The Correctness Problem

Unfortunately, asynchronous circuits are difficult to design because they are concurrent systems. When trying to understand a synchronous circuit, it is possible to pretend that events occur in lock-step. Variations in the speeds of functional elements can be ignored, since the clock is timed to allow for the worst-case delays. The lock-step model does not work for speed-independent designs — a correct design must exhibit proper behavior no matter what the speeds of its components. When several components are operating concurrently, there may be a very large number of possible execution paths, each corresponding to a different set of delays in the components. Nondeterminism resulting from unknown or varying delays is the essential problem in all concurrent systems. For a property to hold for a concurrent system, it must hold for every possible execution.

The usual practice in conventional programming is to implement a prototype, then run it on test cases and debug it until it seems to be correct. This is a hazardous approach for conventional programs if it is important that they be bug-free. It is even more dangerous for concurrent systems, because only a small fraction of the potential executions can be tested. Simulation has the same deficiency; irreproducible errors will likely occur in the system at a later stage in development or use. This is particularly troublesome in VLSI, because the costs of repairing an error at a later stage in the design or processing of the circuit can be exceedingly expensive, particularly if the chip has already been manufactured and sold.

One solution to the problem is to *verify* that a circuit design is correct. Unlike testing or simulation, verification provides an *assurance* that a design is correct. Manual verification is often quite tedious, so it is desirable to automate the process.

### 1.1.3. Scope of the Thesis
The thesis develops and implements a theory for practical automatic verification of speed-independent control circuits. This requires developing a formal model of circuit operation, defining the proper relationship between an implementation and its specification, and constructing a computer program that can check this relationship. The boundaries of this investigation have been chosen to focus on a set of closely related issues and to limit the scope of the project.

First, only speed-independent circuits are considered. Not all asynchronous circuits are speed-independent — some depend on assumptions about the relative timing of components. However, speed-independent circuits are the most interesting as a potential solution to the timing problems of VLSI. Furthermore, speed-independence is a special case of timing-dependent circuits; we have chosen to solve the simpler speed-independent verification problem before attacking even more difficult problems.

Another restriction is that the thesis is primarily concerned with *control circuits*. The appropriate formalisms and methods for the design and analysis of, say, a cache controller are markedly different from those for a combinational multiplier circuit. The important aspect of the former is the sequencing of its input and output signals, but the primary interest in the latter is its effect on data; this is the major distinction between *control* and a *data*. The verification problems for these two types of circuits are also different. Control circuits require reasoning about sequencing; data circuits require reasoning about values. The problem of verifying asynchronous data circuits is not substantially different from verifying synchronous circuits, a problem that has been extensively studied [3,5,18,35].

Verifying control circuits is more closely related to the interesting and currently less tractable problem of verifying concurrent systems. The idea that hardware designs should be separated into control and data parts [82] is gaining acceptance, so it is reasonable to hope that the control parts will be separately verifiable, too.

We consider circuits at the *gate and element level* of abstraction. At this level, it is reasonable to distinguish between inputs and outputs, and to consider data transmitted on individual wires. A somewhat lower level of abstraction would be the *switch level*, which models MOS transistors as bilateral devices (but still manipulating digital, not analog, signals). At a somewhat higher level, it would be appropriate to consider data in units of bytes or words, and to consider the encodings of higher-level data types (such as integers).

Finally, there are two levels of formalism that should be distinguished in hardware specification or description languages: the user languages and the underlying models of behavior upon which they are founded. This research concentrates on a single formalism (trace theory) which can act as the foundation for a variety of different user languages. Although we make use of an ad hoc user language in Chapter 5 in order to present examples, an in-depth exploration of alternative user languages and their merits and deficiencies is beyond the scope of the thesis.

## 1.2.  Verification Issues

There are a number of important questions which should be addressed in any theory of verification.

An obviously crucial issue is the meaning of verification. It is generally only meaningful to consider correctness *relative to a specification*. The relationship between a specification and verification should allow an implementation to *exceed* the minimum requirements stated in a specification. This is a familiar idea in other domains, also: if "regular" gasoline is specified for an automobile, "premium" gasoline may be used instead; if an analog circuit design specifies a 10% resistor, a 5% resistor may be substituted.

Another question is: what type of properties are to be modeled and verified? In concurrent program verification, a distinction is usually made between *safety properties* and *liveness properties*. A safety property asserts that "nothing bad happens", while a liveness property asserts "something good happens." An example of a safety property is "no more than one user accesses the resource at any time." An example of a liveness property is "if the customer waits in line long enough, he will eventually be served." Generally, safety properties can be expressed as conditions on finite computations while liveness properties are conditions on the indefinite future. It is important to be able to verify both safety and liveness properties of asynchronous circuits. Common safety and liveness properties for circuits are: "an input never changes unless the circuit is ready for it," and "if an input changes, a particular output will eventually change." If liveness properties are not verified, circuit designs may deadlock when they should not. Models of processes must capture both liveness and safety aspects of behavior, and specifications must be able to state safety and liveness requirements. Many existing models either do not capture liveness properties, or do so in only a limited way.

An important factor in any model of concurrency is the model of inter-process communication. For example, shared variables and message passing through buffered or unbuffered channels are widely-used models. In asynchronous circuits, communication occurs over individual wires. A wire can connect at most one output together with any number of inputs. Wires cannot store more than one value at a time. At this level, we do not want to require that any particular communication protocol be used — anything should be allowed as long as it ensures the correct operation of the circuit. Hence, message-passing is an inappropriate model because it necessarily involves lower-level communication for synchronization (either between the processes themselves or between the processes and the channel). Wires *can* be regarded as shared variables, but with the restriction that at most one process (circuit) can write to the variable.

A formal model of circuit behavior needs some way to express restrictions on inputs. We would like to be able to model the signals on wires as *logical values*, not continuous quantities. However, the digital model is sound only if restrictions on the use of a circuit are accepted. If the circuit inputs change at an awkward time, there can be behavior that violates the digital abstraction (for example, pulses or voltages at levels between logic thresholds).

The central *practical* issue is *computational feasibility*. We are lucky that the systems we consider are finite-state — perhaps the problem is decidable. Nevertheless, it seems to be inherently difficult. Non-trivial questions about digital circuit properties are almost always NP-hard, and many problems involving finite-state automata are PSPACE-complete. Therefore, it is imperative that a verification methodology support *modular* and *hierarchical* verification. A system is *modular* when it can be described as a collection of modules with limited, well-defined interfaces. The description of the interface should be sufficient to assure that the module will function correctly in the system, without further reference to the internals. In this way, the complexity of the system can be controlled. In considering any part of the system, irrelevant details about the other parts can be suppressed by hiding them behind the interfaces. *Hierarchical* verification means that specifications at a low level of abstraction can be used as descriptions of primitives at the next level. Hierarchical verification allows the task to be broken into smaller parts.

Finally, when does verification happen? Usually, verification is regarded as something that happens after a system has been completely designed (*a posteriori* verification).

However, modular and hierarchical verification can allow verification to proceed in conjunction with system design. For example, in a top-down design each refinement can be verified as it is developed.

## 1.3.  Background and Related Work

This thesis touches directly or indirectly on a large number of topics. This section surveys some of the most relevant work: correct synthesis as an alternative to verification and methods of modeling and verification of concurrent systems which could possibly be adapted to asynchronous circuit verification.

### 1.3.1. Asynchronous Circuit Synthesis

Verification is not the only way to assure correctness. One possible alternative is to design a correct circuit in the first place. More precisely, a specification could potentially be transformed into an implementation using rules or procedures which guarantee that correctness is preserved. Before developing an extensive theory of verification, we should first examine whether sound synthesis could be used instead.

It is sometimes best if large circuits are implemented in a *distributed* way using *standard* components. A distributed implementation is explicitly subdivided into small communicating processes which communicate over well-defined interfaces, with a layout reflecting this structure. In any large circuit, there will be non-trivial communication costs between physically distant components. A distributed implementation takes these costs into account in the design, so that any ill effects on correctness or performance can be minimized.

Standard components are subcircuits which can be used in many different designs and in many places in the same design. By using standard components instead of custom-designing, the investment in finding an especially good design can be amortized over all its uses. This usually saves design time and results in better circuits (because they use faster primitives). We leave open the questions of the extent to which a standard design will be used, and what design decisions are bound in the design (for example, transistor sizing or layout may not be specified).

Almost all of the work (particularly early work) in asynchronous circuit synthesis is *state machine synthesis*: a state machine specification is implemented as a combinational network with delayed feedback. This includes Huffman circuits [45], almost all of

Ungar's book [78], most textbooks in logic design that discuss asynchronous circuits [33,37], and more recent work [36,57,72,73]. These methods either are restricted in the class of circuits they can implement, or they do not always lead to an efficient implementation. The general methods require arbitration inside the circuits to prevent hazards, but arbitration is expensive and it is not known how to optimize its use.

Another problem is that state machines are not modular, so state machine synthesis does not help with finding a distributed implementation of a specification (there are some methods for state machine decomposition, but they are not practical even for synchronous circuits). A better idea is to implement the processes of a distributed specification as state machines; however, this method does not say how to incorporate standard circuits into the design. In general, finding a good design using state-machine synthesis is very difficult, so the use of standard components is important.

In one of the original papers on asynchronous circuit design, Muller defined speed-independent circuits and identified a subclass called *semi-modular* circuits [53,58]. (The intuition behind semi-modularity is that an input change cannot disable an enabled output in any component of a circuit.) By adhering to certain rules, it is possible to generate circuit designs that are guaranteed to be semi-modular. There are two reasons that these results do not lead directly to an adequate synthesis procedure. First, Muller's definition of speed-independence does not include circuits that make nondeterministic output choices, such as mutual-exclusion elements. This is a serious problem, since mutual exclusion is arguably *the* central issue in concurrent systems. Second, although the generating rules ensure that a circuit will be well-formed in some ways, there is no concept of deriving a circuit that meets a particular specification.

Patil has proposed implementing a Petri net specification (see below) of an asynchronous circuit directly in a *asynchronous logic array*, which simulates the places and transitions of the net, in a standard layout analogous to a PLA. His implementations are not speed-independent. Furthermore, they are quite inefficient, since they have distinct modules for every place and transition in the Petri net.

Clark, Molnar and others have implemented *macromodules*, which are delay-insensitive data and control elements [28]. Macromodules are modular, speed-independent, and general (general-purpose computers have been implemented using them). However, macromodules are higher-level than the circuits we would like to consider: using them has more in common with writing a program than designing a circuit. This precludes lower-level circuit optimization. The correctness problem is addressed by providing a set of components that are easy to work with; there is no inherent notion of specification and implementation.

Martin has proposed a way of refining CSP-like programs to self-timed circuits, using correct transformations. He has derived a distributed implementation of the mutual-exclusion problem using this method [49]. He has also published an implementation of a fair mutual-exclusion element that uses an unfair mutual-exclusion element internally [50]. Martin's method is manual*, however there are several efforts under way to automate the translation from CSP-like languages to self-timed circuits [17,79].

Only the last technique meets the requirements of producing a distributed implementation using standard components. It has not been established that circuits resulting from these methods are as good as those designed at least partially by hand. Currently, it seems that the most attractive possibility would be a hybrid system: a transformational synthesis technique which allows a designer to try creative refinements (which are not necessarily allowed by pre-defined transformation rules), then verify that the results are correct. In essence, this would allow a designer to invent new transformations on the fly and still be assured of the correctness of the design.

Even if a transformational synthesis technique could produce optimal circuits, there remain three verification problems: showing that the specification is correct (by comparing with other specifications), showing that the transformations are correct, and showing the the standard circuits are correct. Automatic verification could apply to these problems, also.

### 1.3.2. Models of Concurrency

A theory of verification requires a formal model of the objects being verified. For modularity, such a semantics should be *syntax directed:* the meaning of a construct should be defined in terms of the meanings of its parts. Here we consider some of the existing models of concurrency with these properties.

*Trace Models*

Hoare [39] has proposed modeling the behaviors of concurrent processes as sets of sequences, called *traces.* The elements of the sequences are possible communications with other processes. Trace semantics is appealing because it is simple. Furthermore, if the trace sets of processes are regular, they can be represented as finite automata, which are very well understood and easy to analyze. Unfortunately, trace models are inadequate for representing some common features of programming languages involving certain types of nondeterministic behavior.

---

*It has been automated since this was written.

CCS ("Calculus of Communicating Systems") is a well-known model of concurrency due to Milner [54], which models behaviors as *trees*. CCS has a nice set of operations for composing behaviors and hiding events in them, with accompanying algebraic laws. Milner defines a relation on computation trees called *observational equivalence*, which holds when they cannot be distinguished by external observation. Observational equivalence has been proposed as a basis for verification [13], but it does not satisfy our criterion that an implementation should be allowed to exceed the minimum requirements of a specification. A disadvantage of CCS is that trees are much more difficult to deal with than sequences. This is illustrated by the complexity of the definition of observational equivalence.

Hoare, Brookes, and Roscoe have proposed a formal semantics for CSP in which behaviors are finite sequences terminating with a set of sets of communications which can be refused, called *failures* [14,40,41]. If two processes have the same sets, they are said to be *failure equivalent*. Failure equivalence is less strict than Milner's observational equivalence, but more strict than trace equivalence. Similarly, it is intermediate in mathematical complexity. Many other notions of equivalence for CCS have been proposed, also.

Communication in CSP and CCS is by *synchronization*: a communication occurs when the two processes involved both decide to do it (unbuffered message passing). It is not clear how to adapt these models for wire communication in low-level asynchronous circuits.

### 1.3.3. Temporal Logic

Temporal logic [48,61] is a modal logic tailored for reasoning about situations that change over time. Temporal logic augments propositional or first order logic with a number of modal operators. For example, if $f$ and $g$ are formulas in the logic, the formula $\Box f$ says that $f$ holds at all future times, $\Diamond f$ says that $f$ holds at some future time, and $f \, \mathcal{U} \, g$ says that $f$ is true until $g$ becomes true. Many other modal operators and variations on these operations have been proposed [4,83]. Temporal logic can express both safety and liveness properties. For example, $\Box(P \vee Q)$ ("$P$ or $Q$ is always true") is a safety property, and $\Box(P \rightarrow \Diamond Q)$ ("If $P$, eventually $Q$") is a liveness property.

For low-level hardware verification, there is a close correspondence between logical signals on wires and truth values of propositions, so propositional temporal logic is very appropriate. An added advantage is that propositional temporal logic is decidable. The most promising decision procedure is the translates a temporal logic formula into a finite automaton on infinite sequences[71].

*Axiomatic Approach.*

One approach to hardware (and program) verification is to write axioms for the behaviors of the primitives (for example, gates, flip-flops, and C-elements), then show that
these axioms imply the specification. We call this the *axiomatic approach.* The implication can be proved by applying an appropriate set of axioms and inference rules.
For low-level reasoning about digital circuits, it is reasonable to consider all variables to
be binary, so it is possible to use *propositional temporal logic.* A number of individu-
als have applied linear-time temporal logic to asynchronous circuits, using an axiomatic
approach to verification.

Using first-order linear-time temporal logic, Malachi and Owicki have characterized
and extended correctness conditions, originally stated by Seitz [47,69], for a class of
speed-independent circuits. This work does not address the same problem as this thesis
(verifying particular designs) because it only specifies the properties without proving
them, and because the properties are design rules for a *class* of circuits, not correctness
conditions for a single design.

Bochmann applied propositional temporal logic to the verification of a particular self-
timed circuit [11]. He demonstrated a bug in a published arbiter design [70], and showed
that a modified design was correct. The axioms for the circuit elements did not model
gate delays. This simplification made his proof easier (it was done manually); however,
it caused a bug in his "corrected" design to be overlooked [29].

Mark Bennett has also applied linear-time temporal logic to asynchronous circuit verification, in his PhD Thesis [6]. Generally, he uses an axiomatic approach with linear-time
propositional temporal logic, although he pre-processes formulas into a kind of state
graph, called an *execution graph.* His examples illustrate the deficiencies of axiomatic,
manual verification. Much of the thesis is devoted to developing helpful notation and
theorems for carrying out proofs, yet the proofs of even the simple circuits he considers
occupy several pages.

It is clear from this work that manual verification of asynchronous circuits under speed-
independent assumptions using temporal logic is limited to relatively small circuits. One
alternative which needs to be explored further is the degree to which its applicability
can be extended by using automated decision procedures (Bennett has used a decision
procedure in constructing his execution graphs, but appears not to have applied it directly
to the problem).

*Model Checking Approach*

An alternative to the axiomatic approach is to use *satisfaction* instead of *implication* as the basis for verification. The general method is to check whether a model of the behavior of the circuit satisfies a specification written as logical formulas.

Most of the work in this section is based on CTL (computation tree logic), which is a *branching-time temporal logic* [27]. (In branching-time temporal logic, models of formulas are trees instead of sequences.) CTL has the advantage that model checking is possible in polynomial time in the sizes of both the logic formula and the state graph (and, in fact, is very fast in practice). This procedure is embodied in the CTL model checker, a program which takes a finite state graph representing the program (called a *Kripke structure*) and a formula in CTL, and reports whether the Kripke structure satisfies the formula. It is possible to represent live and fair behavior in the Kripke structure by stating properties which must be true infinitely often in any path through the state graph.

Bugs have been discovered in a number of previously-published, non-trivial asynchronous circuits using model checking, and modified designs have been shown correct. Examples include a self-timed queue cell [15,55]; the arbiter verified by Bochmann (a bug was discovered in his verified design [29]); and a patented queue element [16].

Finally, there are a few cases in which a model-checking approach is applied using *linear* temporal logic instead of CTL. First, Bennett's execution graphs (see above) are something like Kripke structures, so perhaps that approach could be called model checking. Also, Fujita et al. have implemented a verifier for *synchronous* circuits in Prolog [34]. They have a program which translates a variety of representations, including program and gate-level descriptions, into a state graph. This state graph can be compared with linear temporal logic formulas using a Prolog program. They have verified at least one circuit of substantial complexity. Fujita's program incorporates so many tricks that it is difficult to tell which are the most helpful, or get an impression of its overall efficiency given the examples that are provided.

Compositional operations have not been completely worked out for Kripke structures (particularly when there are fairness constraints). Mishra has defined a hiding operation on Kripke structures, called *restriction* [55]. The operation was quite complicated, primarily because of the tree semantics of CTL.

The model checking approach has the advantage of being automatic. One current limitation in dealing with large circuits is that the Kripke structure for the whole circuit must be completely constructed before it can be checked. Sometimes these state graphs are very large, but only small portions of them need to be examined to detect an error.

### 1.3.4. Modular Verification

Because verification is inherently computationally expensive, it is important that it be modular. It should be possible to verify parts of a system independently, then combine the verified properties to obtain a property that is true of the whole system (without re-verifying). Hoare logic (in sequential programming) is modular: pre- and post-conditions can be verified without referring to the context, and can be used to verify properties of the larger program. Preconditions allow assumptions about the environment of a program fragment to be included in a specification, without knowing all about the environment.

Misra and Chandy have proposed a modular method for verifying safety properties [56]. Their model of process communication is based on CSP. Assertions are of the form $r|h|s$, meaning that if $s$ holds initially in $h$ and if $r$ holds for any trace up to step $i$, $s$ will hold in that trace up to step $i + 1$. This method is applicable only to safety properties. This work has been extended to handle additional language features by Zwiers et al. [85].

Barringer, Kuiper, and Pnueli have proposed a modular way of using temporal logic to verify processes with shared variables [4]. They introduce *edge propositions* in logical formulas in order to distinguish between process actions and environment actions. This system, particularly the crucial rule for parallel composition, is very complicated and would probably not be practical even if partially automated. It is not clear how to separate the implicit assumptions about the environment from the assumptions about the process itself.

Pnueli has proposed a simpler way to distinguish between process actions and environment actions, by restricting process composition: a collection of processes communicate using *distributed variables* if each variable can be written by no more than one process. Pnueli proposes two different schemes. In the first, a notation is used similar to Hoare's and Misra and Chandy's, which has separate assertions for the environment and process [62]. Unfortunately, his parallel composition rule is sound only for specifications of safety properties. Pnueli also proposes a notation in which a single assertion is used for each process (also assuming distributed variables); this method is sound.

The distributed variables assumption is reminiscent of low-level communication among asynchronous circuits (outputs of circuits cannot be connected together). Assumptions about the environment are a nice feature; perhaps specifications could be simplified by restricting the contexts in which a process can be used. Also, we have already noted the need for restrictions on environments in *models*, in order to maintain the validity of the digital abstraction.

### 1.3.5. I/O Automata

Another modular verification method uses *I/O automata* [46]. I/O automata are similar to finite automata, except that they may have an infinite number of states and a distinction is made between input and output signals. There are additional conditions on the automata which are intended to ensure that inputs are allowed to change freely: there must be a transition out of every state on every input symbol.

The question of whether a specification constrains input signals concerns us greatly. However, the solution to the problem adopted by I/O automata is not completely satisfactory, because it is a property of the *structure* of the automaton, not the formal language accepted by the automaton. There can be two equivalent automata (which thus express the same specification), one of which meets the condition and one of which does not. We would like to characterize the desired property for formal languages, not automata, so that when two automata have the same formal language, they are both legal or both illegal.

I/O automata have been used to prove the correctness some very difficult distributed algorithms. The general method is to define a simulation relation between an automaton representing detailed behavior (the implementation, in our terminology) and more abstract behavior (the specification). Proofs using I/O automata have not been automated. For our task, we would prefer a less powerful formalism that is easier to automate.

### 1.3.6. Trace Theory

Although trace models of CSP were discussed earlier in this section, trace *theory* is discussed separately here because it is a distinct body of work. Trace theory for speed-independent VLSI circuits was proposed by Rem, Snepscheut, and Udding in 1983 [66,73]. Circuit behaviors are represented as sets of traces, which are sequences of transitions. In their model, a shared action occurs in a composition of two processes when both processes executed it simultaneously (as in CSP).

Snepscheut [72,73] considered trace structures with inputs and outputs (so-called *directed trace structures*). Whenever an output is connected to an input, Snepscheut's *agglutination* (also called *r-composition*) places an implicit delay on the wire between the two. The delays could store any number of input transitions before producing output transitions. Physical wires cannot store an unbounded number of symbols, so agglutination gives a realistic model only when circuits send no more than one signal on a wire at a time (when several transitions are sent through a wire before any emerge from the other end, it is called *transmission interference*). These delays resulted in trace sets which were not regular. Snepscheut proposed that trace specifications be implemented using state machine synthesis.

Udding [76,77] has investigated *delay-insensitive circuits*, which retain the same behavior when delays are added to their input and output wires. He states several sets of
conditions which are sufficient for delay-insensitivity. His composition operator does not
place delays on wires, but there are severe restrictions on its applicability.

Neither Snepscheut nor Udding could capture liveness properties in their models.
Black [9] extends trace theory to sets of infinite traces (represented as Muller automata)
in order to capture liveness and fairness properties. He defines a composition operator
for complete traces which is similar to Snepscheut's and Udding's, in that it places delays
on all wires.

Of the work on trace theory, Ebergen's comes the closest to being suitable for verification [30,31]. Ebergen defines a composition operation on directed trace structures
which does not add delays between inputs and outputs, and which produces reasonable
results without harsh composability restrictions. He also gives a test for whether a composition of two trace structures meets a specification given by a third trace structure.
Like Snepscheut and Udding, he considers only prefix-closed sets of finite traces, so he
cannot model liveness and fairness properties.

Trace theory is appealing because regular sets are so well understood, and because
there are potentially nice compositional properties. Except for Ebergen, who provides a
test without a rationale, none of this work defines what it means for an implementation
to meet a specification. Even in Ebergen's case, there are no suggestions for how
to demonstrate the property. Only Ebergen gives a realistic and general composition
operation (but not for infinite traces), and no one has proposed an operation for hiding
signals in a *directed* trace structure. Also, although it would appear that some of this
work could be automated, no one has proposed doing so.

### 1.3.7. Petri Nets

*Petri nets* are a widely used model of concurrent systems [60]. Structurally, a Petri
net is a directed graph with two types of nodes: *places* and *transitions*. At any time a
Petri net has a *marking*, which is an assignment of some number of *tokens* to its places.
The marking changes when transitions *fire*. A transition is enabled to fire when all of
its input places have one or more tokens in them; when it fires it removes these tokens
and adds new tokens to all its output places. In general, the "state" of the system at any
time is represented by the marking.

There is a dauntingly large body of literature about Petri nets, identifying properties of
markings, subclasses of nets, decision procedures for properties, definitional variations
and extensions of every kind, and applications in system specification, modeling, and
performance analysis. Petri nets of different types have been used as specifications for

asynchronous circuits [25,52,57,59,68]. They have also been used extensively in protocol verification [8,12].

Empirically, Petri nets provide a very clear and succinct way to specify or describe some circuits. If a Petri net has an upper bound on the number of tokens that can appear in a marking (in which case it is said to be *bounded*), it can be regarded as a regular sequence of transition firings. So bounded Petri nets can be used as a notation for trace structures.

A potential of advantage of Petri nets is that some properties can be determined in sub-exponential time. For example, a *place invariant* is a linear relation among the number of tokens in a subset of the places of a system. It has been suggested that this information can be used to prove the presence or absence of anomalies in a system. These ideas have been used in at least one example of protocol verification [7]. However, the relationship between invariants and more interesting anomalies is not clear in the general case; most analysis programs work by building a tree of all reachable markings.

### 1.3.8. Reachability Analysis

Reachability analysis involves constructing a global state graph representing the combined behavior of a set of concurrent processes modeled as cooperating finite-state machines, then inspecting this graph for anomalies. In much of this work, verification is interpreted to mean checking for some *standard* property: absence of deadlock, liveness, stability, and so on. The meaningfulness of these properties generally depends on the interpretation of the formal model being used and on the application, in contrast with the view we have taken that correctness is meaningful only with respect to some specification. Although reachability analysis can demonstrate some failures of some standard liveness properties under a some computational models, it cannot handle arbitrary liveness properties, because it does not consider infinite behaviors.

Reachability analysis is a commonly-used technique for protocol verification [10,84] (it is sometimes called *perturbation analysis*). Holzmann [42,43] discusses ways to control the state explosion inherent in these techniques, including the use of heuristic search of the global state graph to find problems without having to explore the entire state graph.

Kurshan et al. [1,2] have explored automata on infinite sequences for protocol verification. They have defined a *selection/resolution model*, which represents processes as generalized finite-state automata. The method allows all safety properties and many liveness properties to be checked. Several realistic and quite complex examples have been verified using this system.

Judging from results on real examples, no method appears to deal with significantly harder problems than those of Holzmann and Kurshan, both of which basically search global state graphs.

### 1.3.9. Summary

There are some general observations to be made about the applicability of existing work to our problem.

First, the models of inter-process communication implicit in many formalisms are inappropriate for low-level speed-independent circuits.

Second, there is a need for computer assistance in carrying out proofs, particularly when complicated execution paths are present (as in asynchronous control circuits).

Third, almost all methods for automatically verifying concurrent systems ultimately use state-graph techniques. This includes methods of analyzing communicating finite-state machines, decision procedures for temporal logic, methods based on regular languages, and the reachability construction for Petri nets.

## 1.4.   Results of the Thesis

In short, this thesis presents a version of trace theory which is a uniform basis for formal semantics and verification of speed-independent circuits. Specifically:

*A compositional and modular trace semantics for speed-independent circuits.* Circuit behaviors are represented as *trace structures*, which can be composed. Additionally, output wires can be hidden and wires can be renamed. The semantics correctly models wire communication in low-level asynchronous circuits, and properly expresses restrictions on circuit inputs. There are two types of theories: one captures safety properties (*prefix-closed trace structures*), and the other expresses both safety and liveness properties (*complete trace structures*). All trace structures are finitely representable, and all operations are effective.

*A definition of verification based on safe substitution.* An implementation meets a specification if the former can safely be substituted for the latter in any context. This can be formalized as a binary relation between trace structures which we call *conformation*. This concept is highly advantageous: it allows the same formalism to be used both for program semantics and for specification. The same compositional operations and algebraic laws apply to specification as to models, so the ability to do modular and hierarchical verification is inherited from the semantics. Also inherited is the ability to express restrictions on environments. The system is automatically sound and complete, and conformation is decidable for both prefix-closed and complete trace structures.

*A practical automatic verifier.* A program is described which implements the operations on prefix-closed trace structures and tests the conformation relation. It has discovered a bug in a recently published design, and verified the correctness of an amended design. This example is larger than any that has previously been verified in the published literature (or any unpublished example known to the author). The program has also been applied successfully to a number of other published and unpublished circuit designs. The program also allows interactive application of all operations on trace structures.

*Conformation equivalence.* If two trace structures are conformation-equivalent, they cannot be distinguished in any context *by verification*. This is a weaker equivalence relation than semantic equivalence. The concept of conformation equivalence is probably applicable to other models of concurrency, also.

*A simple definition of and test for delay-insensitivity.* The theory is shown to allow a simple definition of *delay-insensitivity*, including the definition for complete trace structures. The property is decidable, and can be tested by the verifier.

*An application of infinite games.* The property of *receptiveness*, which holds when a trace structure does not control its inputs, is defined for complete trace structures in terms of *infinite games*. The property is shown to be decidable by reduction to Church's solvability problem.

## 1.5. Overview of the Thesis

An attempt has been made to organize this document so that a reader who is not interested in the details of the many proofs can skip them and still understand the important points. Some sections are completely informal, and others consist entirely of lemmas. The remaining sections have some informal discussion of the important issues and results followed by formal statements and proofs of the same results. It is usually not hard to tell where the transition is.

Chapter 2 introduces the basic operations used to construct composite circuits from components. These include the operation **compose**, which connects circuits together by identifying wires with the same name, **hide** which makes some wires unobservable to the external world, and **rename**, which changes wire names. Besides explaining the basic operations, this chapter defines a set of laws of *circuit algebra*. It is proved that any semantic interpretation of the operations must satisfy these laws, or else assign different meanings to two identical circuits. This is proved by showing that there is an algebra of circuit structures which satisfies the laws.

Chapter 3 defines a behavioral interpretation of the operations of circuit algebra which captures the *safety properties* of circuits. In this interpretation, a circuit is represented by a *prefix-closed trace structure*, in which all the trace sets are prefix-closed sets of finite traces. The property of *receptiveness*, which requires that the set of possible traces model all possible inputs, is identified and characterized. A number of typical building-block components for asynchronous circuits are described as examples. Prefix-closed trace structures are shown to be a circuit algebra.

In Chapter 4, the concept of a safe substitution of one trace structure for another is formally defined. When a trace structure $T$ can be safely substituted for $T'$, $T$ is said to *conform to* $T'$. Conformation also induces an interesting equivalence relation between trace structures called *conformation equivalence*, which means intuitively that the structures cannot be distinguished by verification. There are simplifications which can be used to reduce prefix-closed trace structures to a canonical form with respect to conformation equivalence. In the canonical form, the set of failures is redundant — it can be derived from the set of successes. There is a very simple and elegant procedure for deciding whether one canonical prefix-closed trace structure conforms to another. Using the definitions of this chapter, a simple definition of and test for delay-insensitivity is presented. Finally, it is shown that canonical prefix-closed trace structures form a bounded distributive lattice. The meet and join operations in this lattice can be used to decompose specifications and to reason explicitly about environments.

Chapter 5 describes an implementation of the theory presented in Chapters 3 and 4. It is an interactive LISP program which can perform the operations of circuit algebra on prefix-closed trace structures and check conformation. Two example circuits implementing a solution to the *mutual exclusion problem* are verified, each at two levels of abstraction. The program detected a bug in one of the circuits (published in a 1985 VLSI conference). The problem is explained, and a revised version of the circuit is shown to be correct.

Chapter 6 presents mathematical notation and background on infinite sequences and infinite games, for use in the next chapter. There is a discussion of regular languages of infinite strings (called $\omega$-regular languages) and the finite automata that accept them. A form of infinite two-player game is defined. It is shown that for a subclass of these games which is important for the theory of complete traces, the problem of whether the second player can always win is decidable, by reduction to Church's solvability problem.

In Chapter 7, trace structures can be extended to capture liveness and fairness properties, using non-prefix-closed sets of finite and infinite traces. The extended structures are called *complete trace structures*. The theory of complete trace structures parallels that of prefix-closed trace structures, with some variations. The receptiveness property for complete traces is much more subtle than for prefix-closed traces; it is characterized using infinite games. The definition of conformation and conformation equivalence is the same as for prefix-closed trace structures. However, the simplifications to a canonical form for conformation equivalence are more complicated (one of them uses infinite games, also). Unlike prefix-closed trace structures, the set of failure traces is not redundant in the canonical form of complete trace structures. There is a decision procedure for conformation on complete trace structures which is a generalization of the procedure for canonical prefix-closed trace structures. However, it is more complicated and computationally more difficult.

Finally, Chapter 8 has a summary of the thesis and some discussion of remaining problems.

# Chapter 2

# Circuit Structure

## 2.1. Introduction

In hardware description languages, a distinction is often made between *structural descriptions* and *behavioral descriptions*. A structural description tells what the parts of a circuit are and how they are connected — it gives the topology of the circuit. A behavioral description tells what the circuit *does*. The purpose of this chapter is to understand circuit structures and how to describe them. In essence, we give a syntax for circuits. In later chapters, we will define behavioral models that are based on the structures of circuits, in the the same way that denotational semantics is based on the syntax of programming languages.

A circuit structure can reasonably be regarded as a kind of graph, essentially a schematic diagram without geometric information. However, this is a flat representation which fails to show regularities and similarities within the structure. In reality, circuits are designed and analyzed *hierarchically*: the circuits are composed of subcircuits which are themselves composed of subcircuits, and so on. Hierarchical designs allow one to focus on individual concerns separately instead of having to grasp the entire operation of the circuit at once. Perhaps more importantly, hierarchical structures give the circuit designer, analyzer, or manufacturer a great deal of leverage; a one-time investment in designing or understanding a subcircuit pays off every time the circuit is replicated. A hierarchical *description* shows how the system is organized.

One way to convert a flat representation into a hierarchical one is to define operations that build large circuits out of smaller ones. A circuit can then be described by an expression using these operations. The structure of the expression shows the hierarchical organization of the circuit. We define the operations **compose, hide**, and **rename**, which operate on circuit structures. **compose** connects circuits together; **hide** makes wires internal, so they cannot be connected to wires of other circuits; and **rename** changes the names of wires. In some cases, different expressions give the same circuit. There is a set of nine algebraic laws that describe exactly when this can happen; two expressions are equivalentequivalence according to the laws if and only if they describe the same circuit.

There are other interpretations of the operations besides the structural one given in this chapter. *Any* interpretation that satisfies the laws is called a *circuit algebra*. In subsequent chapters, various *behavioral interpretations* will be considered. Any reasonable behavioral interpretation must obey the laws of circuit algebra; otherwise, it can assign different behaviors to the same circuit structure.

The operations of circuit algebra are similar to operations in CCS and other models of concurrency. The idea that the algebraic laws might apply to circuit structures as well came from the realization that many identities could be justified simply by drawing block diagrams. Cardelli's thesis [22] also relied heavily on the idea that a single algebra could have a variety of useful interpretations, including a structural one. However, the algebra here is specifically tailored for the description of gate- and element-level circuits, so it is much less ambitious and less general than Cardelli's, and correspondingly less complex. In particular, we do not use many-sorted algebra.

The remaining sections present these results in detail. The second section covers mathematical notation and background for this and subsequent chapters. The third section defines circuit algebra in the abstract: the operations and the nine laws they must satisfy. The fourth section defines a circuit structure as a kind of labeled directed graph. This is a flat representation of circuits. This section defines formally what it means for two circuits to be "the same"; the relation is called *structural equivalence*.

The fifth section gives the structural interpretation: the operations are used to build circuit graphs. This interpretation is shown to be a circuit algebra, which has two important implications. First, since it requires that the set of circuit structures be closed under the operations, every result of the operations is a legitimate circuit structure. Second, if the two expressions are equivalent according to the algebraic laws, they have the same structure. The final section proves the converse: if two circuit expressions are algebraically equivalent, the structures they describe are also equivalent. This result implies that every circuit structure can be described by an expression, and if two circuit structures are equivalent, the expressions describing them are equivalent according to the algebraic laws. The last two results are particularly important for behavioral interpretations: if a behavioral interpretation is *not* a circuit algebra, it either fails to describe some circuits or it assigns different behaviors to circuits having the same structure. Hence, *any behavioral interpretation should be a circuit algebra.* One of the intermediate results in the last section is that every expression in circuit algebra can be reduced to a simple normal form; this result is used repeatedly in subsequent chapters.

For the reader who is prepared to believe the claims made here without detailed proof, Section 2.3 and the first two paragraphs of Section 2.6 are the most important for understanding the rest of the thesis.

## 2.2.  Notation and Other Mathematical Preliminaries

The purpose of this section is to explain some of the notational conventions used in this and subsequent chapters.

Sets are generally written as capital letters. $X \subset Y$ means that $X$ is a *proper* subset of $Y$; if it is a subset but not necessarily a proper subset, the notation $X \subseteq Y$ is used. The power set of $X$ is written $2^X$. The set of non-negative integers (natural numbers) is written $\omega$. The *disjoint union* of two sets $A$ and $B$ is written $A \uplus B$ and is defined to be $A \times \{0\} \cup B \times \{1\}$.

Functions are total, unless otherwise indicated. If a function $f$ has domain $A$ and codomain $B$, this fact is written $f: A \rightarrow B$; the set of such functions is $[A \rightarrow B]$. Occasionally, we use the convention that a natural number is the set of all its predecessors, so $f: i \rightarrow A$ is a function that maps the set $\{j \mid 0 \leq j < i\}$ to $A$. The identity function on $A$ is written $1_A$. Given a function $f: A \rightarrow B$ and $X \subseteq A$, the notation $f(X)$ is used to represent the image of $X$ under $f$, unless $f(X)$ has already been defined in some other way. To avoid confusion in some cases, this notation is explicitly introduced by saying that "$f$ is naturally extended to sets." If $Y \subseteq B, f^{-1}(Y)$ is the inverse image of $Y$ under $f$. If $f$ is a bijection and $a \in A, f^{-1}$ is the inverse of $f$. Otherwise, $f^{-1}(a) = f^{-1}(\{a\})$. It is useful to know certain properties of image functions. If $f$ is a function from $A$ to $B$ and $X, Y \subseteq A$, then the following properties hold:

(2.1) $$f(X \cup Y) = f(X) \cup f(Y),$$

(2.2) $$f(X \cap Y) \subseteq f(X) \cap f(Y),$$

(2.3) $$f^{-1}(X) \cup f^{-1}(Y) = f^{-1}(X \cup Y) \quad \text{and,}$$

(2.4) $$f^{-1}(X) \cap f^{-1}(Y) = f^{-1}(X \cap Y).$$

If $A' \subseteq A$, the *restriction of $f$ to $A'$* is written $f|_{A'}$; this function is the mapping from $A'$ to $B$ having $f|_{A'}(a) = f(a)$ for all $a \in A'$. Sometimes, both the domain and the codomain of $f$ need to be restricted. If $A' \subseteq A$ and $B' \subseteq f(B), f|_{A' \rightarrow B'}$ denotes the mapping from $A'$ to $B'$ which has $f|_{A' \rightarrow B'}(a) = f(a)$ whenever $a \in A'$ and $f(a) \in B'$. If $f(a) \notin B'$, $f|_{A' \rightarrow B'}(a) = f(a)$ is undefined (however, this construct is only used when $f(A') \subseteq B'$).

There are a number of mathematical objects in this document which consist of sub-parts, with some associated abbreviations of the subparts. In order to reduce the tedium inherent in introducing these names over and over again, we define the general structure and give "standard names" for it, its parts, and the related abbreviations. The standard names for the parts generally inherit subscripts and superscripts. For example, we could define a university $U$ to be a triple $(A, F, S)$, representing the set of administrators, faculty, and students. The set $E$ of university employees could be defined to be $A \cup F$. Then, if $U'$ is a university, $A'$, $F'$, $S'$, and $E'$ are automatically assumed to be the administrators, faculty, students, and employees of $U'$.

## 2.3. Circuit Algebra

A *circuit algebra* is defined on a set of *circuit descriptions* (more briefly, *circuits*). We assume the existence of an infinite universal set of wire names, denoted by *Wires*. Every circuit description $C$ has associated with it a finite set $I \subset$ *Wires* of *input wires* and a finite set $O \subset$ *Wires* of *output wires*. $I$ and $O$ must be disjoint. For brevity, we define $A$ to be $I \cup O$; it is not always necessary to say whether a member of $A$ is an input or an output, in which case it is just called a *wire*. A circuit algebra also has three operations: **compose** (abbreviated $\|$), **rename**, and **hide**.

Composition combines two circuits into a single circuit by identifying wires which have the same name. The composition $C'' = C \| C'$ is defined when $O \cap O' = \emptyset$. "Wired-ORs" and tri-state outputs cannot be modeled directly; however, most circuits containing them can be translated to equivalent circuits having only disjoint outputs. The wires of $C''$ are defined so that $O'' = O \cup O'$ and $I'' = (I \cup I') - O''$. This definition ensures that even after many successive compositions, no two outputs will be connected together (although many inputs may be connected). Figure 2.1 is a before-and-after depiction of the composition of two circuits.

In a circuit algebra, composition is associative and commutative:

C1 $$(C \| C') \| C'' = C \| (C' \| C'') = C \| C' \| C''$$
C2 $$C \| C' = C' \| C$$

The operation **rename** changes wire names. This operation takes a *renaming function*, **r**, which is a mapping from "old" wire names to a set of "new" names. $C' = \textbf{rename}(\textbf{r})(C)$ is defined when $\textbf{r} \in [A \to A']$ is a bijection. The resulting circuit has $O' = \textbf{r}(O)$ and $I' = \textbf{r}(I)$

The **rename** operation has several more algebraic laws.

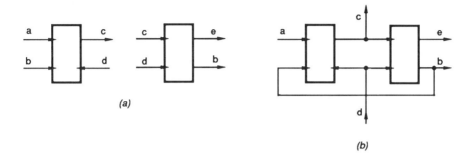

**Figure 2.1.** Composition of Circuits

C3                         **rename(r)[rename(r′)(𝒞)] = rename(r ∘ r′)(𝒞).**

Also,

C4      **rename(r)(𝒞 ∥ 𝒞′) = rename(r|<sub>A→r(A)</sub>)(𝒞) ∥ rename(r|<sub>A′→r(A′)</sub>)(𝒞′),**    and

C5        **rename(1<sub>A</sub>)(𝒞) = 𝒞**

In property C4, both the domains and ranges of the renaming functions must be restricted so that the results are bijections. Note that in C4, the right-hand-side may be defined in some cases when the left-hand-side is not; for example, **r** may not be a bijection even though the restricted on the right-hand-side are bijections. However, whenever both sides are defined, they must have the same value.

   **rename** could be extended to use *surjective* renaming functions, so it could be used to identify wires in the same circuit. For example, the inputs of a NAND-gate could be tied together to make an inverter, or an output could be wired to an input to build a feedback loop using only one component (feedback loops through several components are no problem even without the extension). However, this extension would cause serious problems in our behavioral semantics. It does not impose significant restrictions on the types of circuits that can be described: a NAND with inputs tied together can be represented directly as an inverter, and a non-inverting buffer (identity gate) can be used to carry a feedback signal from an output to an input.

The third operation, **hide**, makes wires "internal" to the circuit, so they can no longer be connected to other wires. Formally, $C' = \textbf{hide}(D)(C)$ is defined when $D \subseteq O$. The effect is that $I' = I$ and $O' = O - D$. Figure 2.2 shows the result of hiding the wires $c$ and $e$ in Figure 2.1b. There are several laws relating **hide** to itself and the previous operations. When $D \cap D' = \emptyset$,

C6                              $\textbf{hide}(D)[\textbf{hide}(D')(C)] = \textbf{hide}(D \cup D')(C).$

Also,

C7                                        $\textbf{hide}(\emptyset)(C) = C,$

and if $D \cap A' = \emptyset$ and $D' \cap A = \emptyset$,

C8                       $\textbf{hide}(D)(C) \parallel \textbf{hide}(D')(C') = \textbf{hide}(D \cup D')(C \parallel C').$

Finally, if $\textbf{r} = \textbf{r}'|_{(A-D) \to \textbf{r}'(A-D)}$ and $\textbf{r}(A - D) \cap \textbf{r}'(D) = \emptyset$,

C9                    $\textbf{rename}(\textbf{r})[\textbf{hide}(D)(C)] = \textbf{hide}[\textbf{r}'(D)][\textbf{rename}(\textbf{r}')(C)].$

Note that in C9, the left-hand-side can be defined when the right-hand-side is not, for example when the domain of $\textbf{r}'$ is a strict superset of $A$. When both sides are defined, they must have the same value.

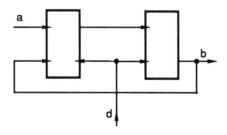

**Figure 2.2.**   Wires $c$ and $e$ Hidden

## 2.4.  A Representation of Circuit Structures

A circuit structure is a collection of primitives, which are called *basic circuits* and two types of wires: internal wires, called *nodes*, and external wires, which are just called *wires*. Every basic circuit has a set of *pins* which are connected to the wires and nodes by a connection function. The same basic circuit can appear more than once in a circuit, so the instances must be distinguished by a set of *basic circuit names*. The nodes have names, too.

In considering whether circuits have the same structure, this representation has certain irrelevant details: it should not matter what the basic circuit names are, nor what the names of the nodes are. Hence, simple equality of circuit structures is too fine a relation. A coarser relation, called *structural equivalence*, is defined which holds if a one-for-one renaming of the basic circuits and nodes can make the structures equal.

The primitives from which a circuit is constructed are the *basic circuits* , the set of which is called *Basic*. A basic circuit $b \in Basic$ has finite, disjoint sets of input and output wires, $i(b)$ and $o(b)$ (both subsets of *Wires*). For brevity, we define $a(b) = i(b) \cup o(b)$.

A circuit structure, $S$, consists of a set of basic circuits connected with wires. Formally, it is a six-tuple, $(I, O, C, c, N, n)$, where $I$ and $O$ are finite, disjoint sets of wire names. As before, $A \triangleq I \cup O$. $C$ is a finite set of *basic circuit instances*. This set is necessary because a circuit may have several instances of the same basic circuit. $C$ is a subset of a universe of potential basic circuit names, *BCNames*, which has the property that $BCNames \times \{0, 1\} \subseteq BCNames$ (because disjoint union is used in composition). The function $c: C \to Basic$ gives the basic circuit corresponding to each basic circuit instance.

Hidden wires in the circuit are represented by a set $N$ of *nodes* . $N$ is a subset of some infinite universe of nodes, written *Nodes*. For technical reasons, we require that $Nodes \cap Wires = \emptyset$ and that $Nodes \times \{0, 1\} \subseteq Nodes$.

A wire of a basic circuit instance is called a *pin*. Formally, a pin is a pair $(c, a)$, where $c \in C$ (a basic circuit instance) and $a \in a[c(c)]$ (the name of a wire in the basic circuit of which $c$ is an instance). $a$ is an input pin if $a \in i[c(c)]$ and an output pin if $a \in o[c(c)]$. The connections between the pins are represented by the *wiring function* $n: \{(c, a) \mid c \in C \wedge a \in a[c(c)]\} \to A \cup N$. If two pins map to the same wire or node, they are connected together. A pin that maps to a wire can, in the future, be connected to pins in other circuits; a pin connected to a node cannot.

It is forbidden for two output pins to be connected, or for a member of $I$ to be connected to an output pin, so we require every member of $n^{-1}(n)$ to be an input pin when $n \in I$, and require $n^{-1}(n)$ to contain exactly one output pin when $n \notin I$. We also require that no two wires of the *same* basic circuit be tied together; formally, for every $c \in C$, we require that $n|_{a[c(c)]}$ be an injection. This restriction is a consequence of the requirement that renaming functions be bijections (see above).

Figure 2.3 shows some schematics that are *not* circuit structures. 2.3a violates the requirement $I \cap O = \emptyset$; 2.3b violates the restriction that members of $I$ (i.e. *a*) cannot be tied to output pins; 2.3c has two pins from the same component connected; and 2.3d has two outputs connected.

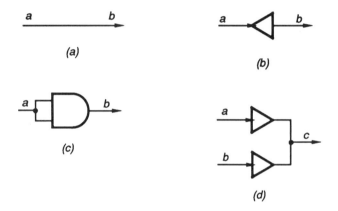

**Figure 2.3.**   Illegal Circuit Structures

Roughly speaking, two circuits are *structurally equivalent* if it is evident from a schematic that they must have the same behavior, even when the behaviors of the components are not known. Also, substituting one circuit for another structurally equivalent circuit in an algebraic expression should give structurally equivalent results.

According to the first criterion, the names of nodes and basic circuit instances do not affect structural equivalence, so long as the connections are the same. More specifically, a one-for-one replacement of the basic circuit instance names or node names should not be considered to change the structure. For example, the circuits in Figure 2.4a and 2.4b obviously have the same structure, but are not equal because their component and node

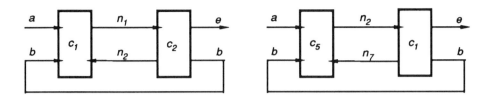

**Figure 2.4.**   Equivalent, but not Equal, Circuit Structures

names are different. This militates for a more liberal notion of equivalence than simple equality of structures (which would require $C = C'$ and $N = N'$ if $S = S'$).

On the other hand, the second criterion (substitutability in expressions) requires that the inputs and outputs of the two circuits match. Otherwise, composition with other circuits could yield markedly different structures.

Formally, circuit structures $S$ and $S'$ are *structurally equivalent* ($S \sim_S S'$) if $I = I'$, $O = O'$, and there exist bijections $f: C \to C'$ and $g: N \to N'$ such that for every $c \in C$, $\mathbf{c}(c) = \mathbf{c}'(f(c))$ (i.e., they have corresponding basic components) and for every $c \in C$ and every $a \in \mathbf{a}[\mathbf{c}(c)]$, either $\mathbf{n}(c,a) \in A$ and $\mathbf{n}(c,a) = \mathbf{n}'(f(c),a)$, or $\mathbf{n}(c,a) \in N$ and $g(\mathbf{n}(c,a)) = \mathbf{n}'(f(c),a)$ (i.e., they are connected in the same way).

The correspondence between the two circuits of Figure 2.4 can be established by having $f$ map $c_1$ to $c_5$ and $c_2$ to $c_1$, and $g$ map $n_1$ to $n_2$ and $n_2$ to $n_7$.

## 2.5.   A Circuit Algebra of Circuit Structures

This section describes an interpretation of the operations **compose**, **rename**, and **hide** in which they build and modify circuit structures. The first definition below does *not* form a circuit algebra, because the representation of circuit structures has irrelevant details that cause problems. However, the operations preserve structural equivalence; that is, if the operands are structurally equivalent, the results will be, also. The interpretation of the operations in which they act on classes of structurally equivalent circuits (the quotient algebra with respect to structural equivalence) *is* a circuit algebra.

This result is important because it implies that the results of the operations are always circuit structures and that if two expressions are algebraically equivalent, the structures they describe are equivalent, too.

When the composition connects two wires, the pins in each structure that were connected to those wires end up being connected to the same wire in the composite. Let $S''$ be the composition $S \parallel S'$. The set, $C''$, of basic circuit names of $S''$, is simply the names from $C$ and $C'$. Unfortunately, we cannot define $C'' = C \cup C'$, because $C$ and $C'$ may have members in common. Instead, we use disjoint union: $C'' \triangleq C \uplus C'$. $\mathbf{c}''$ is defined so that $\mathbf{c}''(c, 0) = \mathbf{c}(c)$ for every $c \in C$ and $\mathbf{c}''(c, 1) = \mathbf{c}'(c)$ for every $c \in C'$. $N''$ is defined to be $N \uplus N'$, and $\mathbf{n}''$ is defined so that $\mathbf{n}''[(c, 0), a] = [\mathbf{n}(c, a), 0]$ when $\mathbf{n}(c, a) \in N$, $\mathbf{n}''[(c, 1), a] = [\mathbf{n}'(c, a), 1]$ when $\mathbf{n}'(c, a) \in N'$, $\mathbf{n}''[(c, 0), a] = \mathbf{n}(c, a)$ when $\mathbf{n}(c, a) \in A$, and $\mathbf{n}''[(c, 1), a] = \mathbf{n}'(c, a)$ when $\mathbf{n}'(c, a) \in A'$. In the first two cases, the original mappings of pins to nodes is preserved (no new connections are formed). In the second two cases, new connections can be formed, because pins from both components may map to a common wire.

If $S' = \mathbf{rename}(\mathbf{r})(S)$, we define $\mathbf{n}'(c, a)$ to be $\mathbf{r}[\mathbf{n}(c, a)]$ if $\mathbf{n}(c, a) \in A$; otherwise $\mathbf{n}'(c, a) = \mathbf{n}(c, a)$. All other aspects of the structure remain unchanged.

The hiding operation on circuit structures removes the hidden wires from the circuit outputs. The hidden wires must be converted into nodes. The same pins remain connected together, but by nodes instead of wires. We assume the existence of an injection $\mathbf{wn}$: $Wires \rightarrowtail Nodes$ to convert wires to nodes (there are many ways to define such a function — any one of them will do). If $S' = \mathbf{hide}(D)(S)$, then $C' = C$, $\mathbf{c}' = \mathbf{c}$, $N' = N \uplus \mathbf{wn}(D)$ (there is no guarantee that $N$ and $\mathbf{wn}(D)$ are disjoint). $\mathbf{n}'(c, a) = [\mathbf{n}(c, a), 0]$ if $\mathbf{n}(c, a) \in N$, $\mathbf{n}'(c, a) = (\mathbf{wn}[\mathbf{n}(c, a)], 1)$ if $\mathbf{n}(c, a) \in D$, and $\mathbf{n}'(c, a) = \mathbf{n}(c, a)$ if $\mathbf{n}(c, a) \in A'$.

The following condition is necessary (but not sufficient) for a circuit algebra:

**Lemma 2.1.** *The set of circuit structures is closed under* **compose**, **rename**, *and* **hide**.

**proof.** Routine. □

Unfortunately, the set of circuit structures is *not* a circuit algebra. In every case, the problem is that disjoint union is neither associative nor commutative. However, if we take the *quotient* of the circuit structures with respect to structural equivalence (i.e., consider the algebra to be equivalence classes of circuit structures instead of individual circuits), the result does indeed obey the laws of circuit algebra. The remainder of this section is devoted to proving these claims.

**Lemma 2.2.** $\sim_S$ *is a congruence with respect to* **compose**, **rename**, *and* **hide**, *as defined above.*

**proof.** The proof considers the individual operations. In the following, let $S_1$, $S_1'$, $S_2$, $S_2'$ be any circuit structures such that $S_1 \sim_S S_1'$ and $S_2 \sim_S S_2'$.

$S_1 \parallel S_2 \sim_S S_1' \parallel S_2'$. Obviously, each composition is defined iff the other is, since the input and output wires are the same. Since $S_1 \sim_S S_1'$, there exist bijections $f_1$ and $g_1$ as in the definition of $\sim_S$. Similarly, there exist bijections $f_2$ and $g_2$ establishing the correspondence between $S_2$ and $S_2'$. Define $f$ so that $f(c, 0) = (f_1(c), 0)$ for all $c \in C_1$

and $f(c, 1) = (f_2(c), 1)$ for all $c \in C_2$. Define $g$ similarly. It is easy to see that $f$ and $g$ are the the required correspondences between $S_1 \parallel S_2$ and $S_1' \parallel S_2'$.

**rename**(r)$(S_1) \sim_S$ **rename**(r)$(S_1')$. The bijections $f$ and $g$ establish a correspondence between **rename**(r)$(S_1)$ and **rename**(r)$(S_1')$ (as well as between $S_1$ and $S_1'$).

**hide**$(D)(S_1) \sim_S$ **hide**$(D)(S_1')$. There exist bijections $f$ and $g$ establishing the correspondence between $S_1$ and $S_1'$. Let $g''(n, 0) = g(n)$ for all $n \in N_1$, and let $g''(n, 1) = (n, 1) = (\mathbf{wn}(a), 1)$ for all $a \in D$. The bijections $f$ and $g''$ give a correspondence between **hide**$(D)(S_1)$ and **hide**$(D)(S_1')$.   $\square$

We denote by $[S]$ the set of all circuit structures that are structurally equivalent to $S$. The fact that $\sim_S$ is a congruence makes it easy to define the operations on equivalence classes of circuit structures:

$$[S] \parallel [S'] \triangleq [S \parallel S']$$

$$\mathbf{rename}(\mathbf{r})([S]) \triangleq [\mathbf{rename}(\mathbf{r})(S)]$$

$$\mathbf{hide}(D)([S]) \triangleq [\mathbf{hide}(D)(S)]$$

We define the *algebra of circuit structures* to be the set of all equivalence classes of circuit structures under these operations.

**Theorem 2.1.** *The algebra of circuit structures is a circuit algebra.*

**proof.** It must be shown that the quotient algebra obeys each of the laws of circuit algebra. In general the proofs are straightforward but tedious, so we give the proof for C8, only. The remaining laws are simpler.

(C8) Let $S$ and $S'$ be any structures for which composition is defined and for which $A \cap D' = A' \cap D = \emptyset$, and let $S_L = \mathbf{hide}(D)(S) \parallel \mathbf{hide}(D')(S')$ and $S_R = \mathbf{hide}(D \cup D')(S \parallel S')$. We are to prove $S_L \sim_S S_R$. Note that $N_L = [N \uplus \mathbf{wn}(D)] \uplus [N' \uplus \mathbf{wn}(D')]$ and $N_R = (N \uplus N') \uplus [\mathbf{wn}(D) \cup \mathbf{wn}(D')]$. Since $\mathbf{hide}(D)(S)$ does not alter $C$, $C_L = C_R = C \uplus C'$; let $f$ be the identity function on this set. Let $g: N_L \to N_R$ be defined so that $g[(n, 0), 0] = [(n, 0), 0]$ when $n \in N$, $g[(n, 1), 0] = (n, 1)$ when $n \in \mathbf{wn}(D)$, $g[(n, 0), 1] = [(n, 1), 0]$ when $n \in N'$, and $g[(n, 1), 1] = (n, 1)$ when $n \in \mathbf{wn}(D')$. Obviously, $f$ and $g$ are bijections (note that $D \cap D' = \emptyset$ by hypothesis).

It is straightforward but tedious to show that for all $n \in N$, $g[\mathbf{n}_L(n)] = \mathbf{n}_R(n)$, by enumerating the kinds of pins of $S_L$ and $S_R$.   $\square$

## 2.6. Structural Completeness

This section proves that if two expressions describe structurally equivalent circuits, the expressions are algebraically equivalent. This implies that any valid behavioral interpretation *must* be a circuit algebra. Otherwise, the interpretation will either fail to describe all circuits, or it will assign different behaviors to structurally equivalent circuits.

One of the important lemmas in the proof of this result is that every circuit algebra expression is equivalent to a *normal-form expression* by the laws of circuit algebra. A normal-form expression consists of a set of innermost **rename** operations composed together, with a **hide** operation applied to them:

$$\textbf{hide}(D)((\dots(\textbf{rename}(r_1)(S_1) \parallel \textbf{rename}(r_2)(S_2)) \parallel \dots \textbf{rename}(r_n)(S_n)\dots)).$$

An *abstract circuit expression* is a sequence of objects: $($, $)$, $\parallel$, **rename**, **hide**, members of *Basic*, bijections from finite subsets of *Wires* to finite subsets of *Wires*, and finite subsets of *Wires*, satisfying the definition below. We follow the usual notation for circuit wires: if $E$ is a basic circuit expression, $I$ is the set of input wires, $O$ the outputs, and $A \triangleq I \cup O$. As before, the input and output wires of the expressions resulting from the operations are completely determined by the definition of circuit algebra, and will not be repeated here.

If $E \in Basic$, $E$ is an abstract circuit expression, $I \triangleq i(E)$, and $O \triangleq o(E)$.

If $E$ and $E'$ are abstract circuit expressions such that $O \cap O' = \emptyset$, the expression of the form $(E \parallel E')$ is also an abstract circuit expression.

If $E$ is an abstract circuit expression and $r$ is a bijection from $A$ to some finite subset of *Wires*, **rename**$(r)(E)$ is also an abstract circuit expression.

If $E$ is an abstract circuit expression and $D \subseteq o(E)$, **hide**$(D)(E)$ is also an abstract circuit expression.

The *free circuit algebra* is constructed by first defining each operation on abstract circuit expressions: the composition of $E_1$ and $E_2$ is simply the expression $(E_1 \parallel E_2)$, and so on. Obviously, the laws of circuit algebra define a congruence relation on abstract circuit expressions; this relation is called *algebraic equivalence*. Hence, it is possible to redefine the operations to act on equivalence classes of circuit expressions. This quotient algebra, which is by definition a circuit algebra, is the free circuit algebra.

A *homomorphism on circuit algebras* is a mapping from one algebra to another that preserves the three operations. Two circuit algebras are *isomorphic* if there exists a homomorphism from one to the other which is a bijection.

**Theorem 2.2.** *The free circuit algebra is isomorphic to the algebra of circuit structures.*

To prove the theorem, we first define the mapping from the free circuit algebra to the algebra of circuit structures. The mapping is defined inductively: If $E$ is a basic circuit $b$, let $\phi(E)$ be the set of circuits structurally equivalent to $(\mathbf{i}(b), \mathbf{o}(b), c, \mathbf{c}, \emptyset, \emptyset)$, where $c$ is any member of *BCNames* and $\mathbf{c}$ is the function that maps $c$ to $b$. In the remaining cases, define $\phi(E_1 \parallel E_2) = \phi(E_1) \parallel \phi(E_2)$, $\phi[\mathbf{rename}(\mathbf{r})(E_1)] = \mathbf{rename}(\mathbf{r})[\phi(E_1)]$, and $\phi[\mathbf{hide}(D)(E_1)] = \mathbf{hide}(D)[\phi(E_1)]$.

We extend $\phi$ to equivalence classes of expressions in the obvious way: Let $[E]$ be the set of expressions algebraically equivalent to $E$. Then $\phi([E]) \triangleq \phi(E)$. $\phi$ is total by Lemma 2.1 and well-defined by Theorem 2.1. $\phi$ is a homomorphism by definition. It remains to be proved that it is a bijection.

**Lemma 2.3.** *$\phi$ is a surjection.*

**proof.** Given any circuit structure $\mathcal{S}$, an expression $E$ can be constructed such that $\mathcal{S} \in \phi(E)$.

The general strategy is to construct the expression bottom-up in three steps. First, for each basic circuit instance, there is an expression: the corresponding basic circuit. Then **rename** and **compose** are used to arrange the connections in the circuit (no wires are hidden, so at this stage the circuit has no nodes). Finally, **hide** is used to convert the hidden wires into nodes. After this step, the expression describes a circuit structure which is structurally equivalent to $\mathcal{S}$.

The first step is: for every $c \in C$, let the expression $E_c = \mathbf{c}(c)$.

For the second step, it is necessary to invent new wire names to represent all the nodes in $\mathcal{S}$. Let $\mathbf{nw}$ be any function which maps every $n \in N$ to a unique wire $a \notin A$, and let the codomain of $\mathbf{nw}$ be $D \triangleq \mathbf{nw}(N)$ (so that $\mathbf{nw}$ is a bijection). We define renaming functions that encode $\mathbf{n}$: for each basic circuit instance, $c \in C$, let $\mathbf{r}_c$ be the function that maps its pins $(c, a)$ to $\mathbf{nw}[\mathbf{n}(c, a)]$ if $\mathbf{n}(c, a) \in N$, or to $\mathbf{n}(c, a)$ if $\mathbf{n}(c, a)$ is a wire (in $A$).

Let the expression $E'_c$ be $\mathbf{rename}(\mathbf{r}_c)(E_c)$. Let $E'$ be the composition of all the $E'_c$ (there are any number of ways to do this, which are all equivalent by C1 and C2, as long as each $E'_c$ is represented in the composite exactly once).

For the final step, the invented wire names ($\mathbf{nw}(N) = D$) must be converted into nodes by hiding. Let $E = \mathbf{hide}[\mathbf{nw}(N)](E')$. $E$ is the desired expression.

To see that $\mathcal{S} \in \phi(E)$, we construct a structure $\mathcal{S}'$ that is definitely in $\phi(E)$, and show that it is structurally equivalent to $\mathcal{S}$.

First, note that for every basic circuit instance $c \in C$, the structure consisting only of that basic circuit, $(\mathbf{i}(b), \mathbf{o}(b), \{c\}, \mathbf{c}|_{\{c\}}, \emptyset, \emptyset)$, is in $\phi(E_c)$. The structure $\mathcal{S}'$ defined by this interpretation of the $E_c$'s is in $\phi(E)$ by the definitions of the operations on circuit structures.

To show $S \in \phi(E)$, we need only show that $S' \sim_S S$.

There is clearly a one-to-one correspondence $f$ between $C$ and $C'$ — the members of $C'$ are the members of $C$, except they have been paired with many 0's and 1's by disjoint unions in composition.

Let $g: N \to N'$ be defined so that $g(n) = \mathbf{nw}^{-1}[\mathbf{wn}^{-1}(n)]$ for every $n \in N$ (the construction uses $\mathbf{nw}$ to map nodes in $S$ to wires in $E'$, and $\mathbf{wn}$ to map these wires to nodes in the hiding step that converts $E'$ to $E$). $f$ and $g$ establish the desired correspondence between $S'$ and $S$.    □

We define the set of *normal form expressions*, $\langle N \rangle$, to be the least set of abstract circuit expressions satisfying:

$$\langle N \rangle \subseteq \mathbf{hide}(D)(\langle C \rangle)$$
$$\langle C \rangle \subseteq \langle R \rangle \cup (\langle C \rangle \parallel \langle R \rangle)$$
$$\langle R \rangle \subseteq \mathbf{rename}(\mathbf{r})(Basic)$$

where each of the operations on expressions has been extended to sets of expressions.

**Lemma 2.4.**  *If $E \in \langle C \rangle$, the expression $E' = \mathbf{rename}(\mathbf{r})(E)$ is equivalent to an expression in $\langle C \rangle$.*

**proof.**  By structural induction on expressions. If $E \in \langle R \rangle$, $E'$ can be converted to an equivalent $E'' \in \langle C \rangle$ by C3.

If $E$ is of the form $(\langle C \rangle \parallel \langle R \rangle)$, $E'$ can be converted to an equivalent $E'' \in \langle C \rangle$ by applying C4, then applying the lemma inductively to the left component and applying C3 to the right component.    □

**Lemma 2.5.**  *If $E = \mathbf{hide}(D)(E_1)$ is a normal form expression and $D'$ is any subset of Wires such that $|D'| = |D|$ and $D' \cap A = \emptyset$, then there exists a normal form expression $E' = \mathbf{hide}(D')(E_1')$ which is equivalent to $E$.*

**proof.**  Let $\mathbf{r}$ be any of the many bijections from $A \cup D$ to $A \cup D'$ such that $\mathbf{r}|_{A \to A} = \mathbf{1}_A$. Then $\mathbf{hide}(D')[\mathbf{rename}(\mathbf{r})(E_1)]$ is equivalent to $E$ by C5 and C9. By Lemma 2.4, this is in turn equivalent to some normal-form expression $\mathbf{hide}(D')(E_1')$.    □

**Lemma 2.6.** *Every expression is algebraically equivalent to an expression in normal form.*

**proof.** Given any abstract circuit expression (regarded as a tree), equivalence-preserving transformations based on the identities can be used to move the **rename** operations towards the leaves and the **hide** operations towards the root.

We prove by structural induction on expressions.

If $E \in$ *Basic*, **hide**$(\emptyset)$[**rename**$(1_A)(E)$] $\in \langle N \rangle$ is equivalent by C7 and C5.

If $E =$ **hide**$(D)(E_1)$, there is a normal form $E'_1$ for $E_1$ by induction. Otherwise, $E \sim_S$ **hide**$(D)(E'_1)$, which can be normalized by applying C6.

If $E = (E_1 \parallel E_2)$, there exist normal forms $E'_1$ and $E'_2$ for $E_1$ and $E_2$ such that $E'_1 =$ **hide**$(D_1)(E_3)$ and $E'_2 =$ **hide**$(D_2)(E_4)$. $E'_1$ and $E'_2$ can be converted to equivalent normal forms $E''_1 =$ **hide**$(D'_1)(E'_3)$ and $E''_2 =$ **hide**$(D'_2)(E'_4)$, where $D'_1 \cap A''_2 = D'_2 \cap A''_1 = \emptyset$ (this is possible by Lemma 2.5). $E''_1 \parallel E''_2$ is equivalent to $E$, and can be converted to the form **hide**$(D'_1 \cup D'_2)(E'_3 \parallel E'_4)$ form by applying C8. This expression can be reduced to normal form by repeated applications of C1 and C2 (to put the compositions in left associative form).

If $E =$ **rename**$(\mathbf{r})(E_1)$, there is a normal form $E'_1 \sim_S E_1$ such that $E'_1 =$ **hide**$(D)(E_2)$. By Lemma 2.5, $E'_1$ has an equivalent normal form $E''_1 =$ **hide**$(D')(E'_2)$ such that $\mathbf{r}(A - D) \cap D' = \emptyset$. Define $\mathbf{r}'$ so that $\mathbf{r}'|_{A-D} = \mathbf{r}$ and $\mathbf{r}'|_D$ is the identity function. $E' =$ **hide**$(D')$[**rename**$(\mathbf{r}')(E'_2)$] is then equivalent to $E$ by C9, and by Lemma 2.4 there is an normal form $E''$ equivalent to $E'$.   □

**Lemma 2.7.** $\phi$ *is an injection.*

**proof.** Let $E$ and $E'$ be any normal form expressions such that $\phi(E) = \phi(E')$. We construct a circuit structure $S \in \phi(E)$. Let $\hat{C}$ be the set of integers $1, 2, \ldots, k$, where $k$ is the number of basic circuit sub-expressions in $E$. Since $E$ is in normal form, we can unambiguously index the subexpressions $E_i \in$ *Basic* with these numbers by the order in which they appear (left-to-right) in $E$. We can index the expressions **rename**$(\mathbf{r}_i)(E_i)$ in the same way. Define $\hat{\mathbf{c}}$ so that $\hat{\mathbf{c}}(i) = E_i$, define $\hat{N}$ to be $D$, and define $\hat{\mathbf{n}}$ to map pairs $(i, a)$, where $1 \le i \le k$ is an integer and $a \in \mathbf{a}[\hat{\mathbf{c}}(i)]$, to $\mathbf{r}_i(a)$.

To define $S$, we first define $C$ by choosing a unique member of *Wires* for each member of $\hat{C}$; let **pc** be the appropriate bijection from $C$ to $\hat{C}$. Define **c** so that $\mathbf{c}[\mathbf{pc}(i)] = \hat{\mathbf{c}}(i)$ for every $i \in \hat{C}$. Define $N = \mathbf{wn}(\hat{N})$ and **n** so that $\mathbf{n}(c, a) = \hat{\mathbf{n}}(i, a)$ when $\hat{\mathbf{n}}(i, a) \in A$ and $\mathbf{n}(c, a) = \mathbf{wn}[\hat{\mathbf{n}}(i, a)]$ when $\hat{\mathbf{n}}(i, a) \in D$. Of course, the inputs and outputs of the circuit structure are the same as those of $E$. Construct $S' \in \phi(E')$ similarly.

$S \sim_S S'$ by the assumption that $\phi(E) = \phi(E')$, so there exist bijections $f$ and $g$ per the definition of $\sim_S$. Define the bijections $f' = \mathbf{pc} \circ f \circ \mathbf{pc}'^{-1}$ and $g' = (\mathbf{wn}|_{D \to N}) \circ g \circ$

$(\mathbf{wn}|_{D' \to N'})^{-1}$. These functions can be used to convert $E$ to $E'$ by equivalence-preserving transformations. Obviously, $f'$ is a permutation of the basic circuit subexpressions, and all such rearrangements are equivalent by laws C1 and C2. $g'$ is a one-for-one substitution of the wire names in $D'$ for those in $D$ which can be done while preserving equivalence by Lemma 2.5. So $E$ and $E'$ are algebraically equivalent.

This result can be generalized to arbitrary abstract circuit expressions by two applications of Lemma 2.6.   □

This completes the proof of Theorem 2.2.

# Chapter 3

# Prefix-closed Trace Structures

## 3.1. Introduction

To verify properties of circuit *behaviors*, we must model them. This chapter models behaviors by defining a *behavioral interpretation* for the operations of circuit algebra. Given an expression describing the structure of a circuit and behavioral descriptions of the basic circuits in it, this interpretation can be used to find a behavior for the whole circuit.

The model is a variant of the trace theory of Rem, Snepscheut, Udding, and others [30,31,66,72,73,76,77]. An execution of a circuit over time is represented as a sequence of events, called a *trace*. The events are *transitions*, which are changes of voltage levels from logical 1 to logical 0 or vice-versa. A transition is represented by the name of the wire on which it occurs. A circuit may exhibit a variety of traces, depending on the sequences of inputs it receives, differences in timing of independent transitions, and arbitrary choices the circuit may make among internal states and output transitions. These alternatives are represented by associating trace sets with the circuit. An entire description of a circuit is called a *trace structure*.

In the introduction, it was noted that there is a fundamental difference between communication in many concurrent programming languages (such as CSP) and communication over wires in low-level circuits. In CSP, a message is not received by another process until that process asks to read it — communication is by *synchronization*, requiring the consent of both parties. In contrast, a circuit cannot control the arrivals of transitions on its inputs, so unwanted inputs can occur. To cope with this, we define a novel property called *receptiveness* which ensures that a trace structure models *all* possible inputs, even the unwanted ones.

Another problem in modeling asynchronous circuits is that the validity of the digital model depends on restrictions on the inputs. Trace structures model such restrictions by having a set of traces called *failures*. An unwanted input is *possible* but is flagged as a bad event by including it in the set of failures.

The trace structures of this chapter are called *prefix-closed trace structures*. The traces represent *partial* executions of a circuit. This representation can express all safety properties, but not liveness properties. We consider prefix-closed trace structures separately from the more general *complete trace structures* of Chapter 7 for two reasons. First, prefix-closed trace structures are much easier to understand. Since the theory of complete trace structures closely parallels the theory of prefix-closed trace structures, it simplifies the presentation to do the simple theory first, then extend on it. Second, the operations and decision procedures for prefix-closed trace structures can be efficiently implemented. This is not (yet) true of complete trace structures. The program described in Chapter 5 is an implementation of the theory of prefix-closed trace structures. Therefore, it can only check safety properties; however, this is very useful since most of the design errors in speed-independent circuits cause violations of safety properties.

The trace theory here is more general and more elegant (in the author's opinion) than the versions of trace theory on which it builds. Additionally, this theory provides a superior basis for verification. Unlike earlier work in trace theory, our theory is not limited to (or particularly concerned with) delay-insensitive circuits. A circuit description is delay-insensitive if it remains the same when arbitrary wire delays are placed on its inputs and outputs. This is a very nice property. However, delay insensitivity is not *essential* — small-scale circuits can often safely assume that wire delays are negligible. Such circuits are often designed under the weaker *speed-independence* assumption, that circuit elements may have arbitrary delays but wires have no delays. In other words, a wire is a single equipotential region [69]. It is thus important to be able to model circuits that are not delay-insensitive.

Although our theory is not limited to delay-insensitive circuits, they can be characterized very simply using it (see Section 4.6). Moreover, specifications can be checked automatically for delay insensitivity using the program of Chapter 5.

In the sections that follow, the second covers mathematical notation and background, relating to sequences and automata. The third section defines prefix-closed trace structures and the operations of circuit algebra on them. The fourth section gives examples of trace structures for common circuits, such as Boolean gates; among other things, these show how input restrictions can be used to exclude hazards and unanticipated metastable states.

## 3.2. Mathematical Preliminaries: Finite Sequences

This section covers mathematical notation and background relating to finite sequences and regular languages, for this and subsequent chapters.

A finite sequence $x$ of length $n \in \omega$ over some set $A$ is a function $\{i \mid 0 \leq i < n\} \rightarrow A$. The length of $x$ is written $|x|$, the $i$th element of $x$ is written $x(i)$. If $A$ does not contain sequences, the set of all finite-length sequences over $A$ is written $A^*$. The sequence of length 0 is the empty sequence and is written $\epsilon$. The *concatenation* of two sequences $x$ and $y$ is the sequence written $x \cdot y$ or $xy$ if no confusion should result. It is defined so that $(xy)(i) = x(i)$ if $i < |x|$ and $(xy)(i) = y(i - |x|)$ if $i \geq |x|$. Concatenation is naturally extended to sets of finite sequences. The concatenation of sets $X$ and $Y$ is written $XY$.

If $X$ is a set of sequences and $i \in \omega$, $X^i$ is defined recursively: $X^0$ is $\{\epsilon\}$ and $X^{i+1} = X \cdot X^i = X^i \cdot X$. $X^*$ is defined to be the union of all $X^i$: $X^* = \bigcup_{i \in \omega} X^i$. $X^+$ is defined to be $X \cdot X^*$.

For sets of sequences $X$ and $Y$, the *quotient* of $X$ and $Y$, written $X/Y$, is defined to be $\{x \mid \exists y \in Y : xy \in X\}$. If $X$, $X'$, and $Y$ are any sets of sequences, quotient has the following properties:

(3.1) $$(X \cap X')/Y \subseteq X/Y \cap X'/Y$$

(3.2) $$(X/Y)/Z = X/(ZY).$$

Quotient preserves set inclusion in its first argument: if $X \subseteq X'$, then $X/Y \subseteq X'/Y$.

The natural extension of a function $f : A \rightarrow A'$ to sequences on $A$ is also written $f$ and is defined recursively so that $f(\epsilon) = \epsilon$ and for any $a \in A$ and $x \in A^*$, $f(ax) = f(a)f(x)$. Note that any function extended in this way distributes over sequence concatenation: $f(xy) = f(x)f(y)$.

A sequence $x$ is a *prefix* of a sequence $y$ (written $x \leq y$) if $|x| \leq |y|$ and $x(i) = y(i)$ for $i < |x|$. It is easy to see that $\leq$ is a partial order. A further property is that if $x \leq z$ and $y \leq z$, then $x \leq y$ or $y \leq x$. If $x$ is a sequence, $\textbf{pref}(x)$ is the set of all prefixes of $x$. Clearly, $\textbf{pref}(x)$ is totally ordered by $\leq$. A set $X$ is said to be *prefix-closed* when $\textbf{pref}(X) \subseteq X$.

We define $\textbf{del}(D)(a)$ for any $a \in A$ and $D \subseteq A$ so that $\textbf{del}(D)(a) = \epsilon$ if $a \in D$ and $\textbf{del}(D)(a) = a$ if $a \notin D$. This function is extended naturally to sequences and sets. We also use the inverse image of $\textbf{del}(D)$, written $\textbf{del}(D)^{-1}(X)$. By the definition of inverse image functions, $\textbf{del}(D)^{-1}(X)$ is the set $\{x' \mid \textbf{del}(D)(x') \in X\}$; i.e., the set of all sequences that would be in $X$ if all $D$ symbols were deleted from them.

There are some simple and useful properties involving these operations on sets of sequences.

(3.3)      $\mathbf{del}(D)[\mathbf{del}(D')(X)] = \mathbf{del}(D \cup D')(X) = \mathbf{del}(D')[\mathbf{del}(D)(X)]$

(3.4)      $\mathbf{del}(D)[\mathbf{del}(D')^{-1}(X)] = \mathbf{del}(D')^{-1}[\mathbf{del}(D)(X)]$      when $D \cap D' = \emptyset$

(3.5)          $\mathbf{del}(D)(X \cap X') \subseteq \mathbf{del}(D)(X) \cap \mathbf{del}(D)(X')$

(3.6)          $\mathbf{del}(D)[\mathbf{pref}(X)] = \mathbf{pref}[\mathbf{del}(D)(X)]$

(3.7)          $\mathbf{del}(D)^{-1}[\mathbf{pref}(X)] = \mathbf{pref}[\mathbf{del}(D)^{-1}(X)]$

(3.8)              $\mathbf{pref}(X \cap X') \subseteq \mathbf{pref}(X) \cap \mathbf{pref}(X')$

We assume familiarity with regular sets and finite automata (as explained in [44]), but reiterate some of the definitions to introduce our particular notation. A *nondeterministic finite automaton*, $\mathcal{M}$, is a five-tuple $(A, Q, Q_F, \mathbf{n}, Q_0)$, where $A$ is a finite set of *symbols* (the *alphabet*), $Q$ is a finite set of *states*, $\mathbf{n}: Q \times A \to 2^Q$ is the *transition function*, $Q_0 \subseteq Q$ is the set of *initial states*, and $Q_F \subseteq Q$ is the set of *final states*. A *deterministic* finite automaton is a modified form of nondeterministic automaton: $\mathbf{n}$ is defined to be a partial function in $Q \times A \to Q$, and there is a single initial state $q_0 \in Q$ instead of the set $Q_0$.

A *run*, $p$, of a nondeterministic automaton on a sequence $x \in A^*$ is a sequence on $Q$, such that $p(0) \in Q_0$ and for all $0 \le i < n$, $p(i+1) \in \mathbf{n}[p(i), x(i)]$ (for a deterministic automaton, $p(0) = q_0$ and $p(i+1) = \mathbf{n}[p(i), x(i)]$). The run is *accepting* if $p(|x|) \in Q_F$. If there exists an accepting run on $x$, $\mathcal{M}$ *accepts* $x$. The *language* of an automaton $\mathcal{M}$, written $\mathcal{L}(\mathcal{M})$, is the set of all sequences accepted by $\mathcal{M}$.

Regular expressions over an alphabet $A$ are defined recursively: every member of $A$ is a regular expression; if $\alpha$ and $\beta$ are regular expressions, so are $(\alpha + \beta)$ (union), $(\alpha \cdot \beta)$ (concatenation, abbreviated $\alpha\beta$), and $(\alpha^*)$ (*Kleene closure*). The usual liberties are taken with syntax: $+$ has weaker precedence than $\cdot$ which is weaker than $^*$, and parenthesis are omitted when the results are unambiguous. $[\alpha]$ is the regular set corresponding to the regular expression $\alpha$.

Regular sets are closed under the union, concatenation, Kleene closure, intersection, complement, quotient with other regular sets, homomorphisms (such as $\mathbf{del}(D)$) and inverse homomorphisms (such as $\mathbf{del}(D)^{-1}$) [44]. It is also easy to see that $\mathbf{pref}$ preserves regular sets.

## 3.3.  Prefix-Closed Trace Structures

This section defines the operations of circuit algebra for trace structures. As usual, the conditions on $I$ and $O$ for definedness and the effects of operations on these sets are completely determined by the earlier description of circuit algebra, and will not be repeated here.

A *prefix-closed trace structure* $T$ is defined to be a four-tuple $(I, O, S, F)$. $I \subset Wires$ is a finite set of *input symbols* (the *input alphabet*), $O \subset Wires$ is a finite set of *output symbols* (the *output alphabet*). We define the abbreviation $A = I \cup O$, the *alphabet* of the structure. Traces in $S$ are said to be *successful traces* (or *successes*); $F$ is the set of *failure traces* (or *failures*). $S$ and $F$ are *not* necessarily disjoint. We define the set $P$ to be $S \cup F$, the set of *possible traces*. $S$ and $F$ (and hence $P$) are regular subsets of $A^*$. (The regularity condition is justified, because circuits have a finite number of internal states.) The sets $S$ and $P$ are also prefix-closed (they represent partial executions). $P$ cannot be empty, since even the circuit that does nothing has the empty sequence ($\epsilon$) as a behavior.

We require trace structures to be *receptive*, meaning that $PI \subseteq P$. Intuitively, a trace structure represents a constraint on the traces that can appear in any composition containing the circuit — the trace $xa$ is possible in a composition if and only if it is in the intersection of the $P$ sets of all the circuits in the composition. If $x \in P$ and $a$ is an input of $T$ ($a \in I$), we must have $xa \in P$; otherwise the circuit could constrain the input $a$ not to be sent.

In essence, the composition of two trace structures has the set of all traces that are consistent with both structures. The alphabet of the composition is the union of the alphabets of its components. Each component constrains the symbols in its alphabet and ignores the others. Composition is defined in two steps. The first step makes the alphabets of all the trace structures the same by adding ignored inputs as necessary to each structure. This definition simplifies some of the proofs that follow.

The inverse of the **del** operation is used in the first step. Intuitively, if $x$ is a trace not containing symbols from $D \subset Wires$, $\mathbf{del}(D)^{-1}(x)$ is the set of all traces that can be generated by inserting members of $D^*$ between consecutive symbols in $x$. Inverse deletion is extended to structures:

$$\mathbf{del}(D)^{-1}(T) \triangleq (I \cup D, O, \mathbf{del}(D)^{-1}(S), \mathbf{del}(D)^{-1}(F)).$$

The wires in $D$ are the additional inputs. This is defined only when $D \cap A = \emptyset$.

For the second step, we want the sets of traces which are consistent with the component trace structures. Set intersection is extended to trace structures:

$$T \cap T' \triangleq [I \cap I', O \cup O', S \cap S', (F \cap P') \cup (P \cap F')],$$

which is defined when $A = A'$ and $O \cap O' = \emptyset$. According to this definition, a trace is a success in the composite if and only if it is a success in both components. A composite trace is a failure if and only if it is failure in either component. Note that the set of possible traces for the composite is $P \cap P'$.

Composition is defined using these two operations:

$$T \parallel T' = \mathbf{del}(A' - A)^{-1}(T) \cap \mathbf{del}(A - A')^{-1}(T').$$

It is illuminating to consider the effect of composition in two boundary cases: when the alphabets of the two traces are disjoint, and when they are the equal. In the first case, the trace set of the composition is the set of *shuffles* (interleavings) of traces from each component trace set. In the second, the trace set of the composite is the *intersection* of the trace sets of the components. Composition finds the set of traces that match at their common symbols and have arbitrary interleavings of the other symbols. Although the definition is somewhat unusual, this composition operation is similar in its essence to most other composition operations. The variations among them are primarily in the treatments of alphabets.

The definition of **rename(r)** is simply the natural extension of **r** to sequences and then sets. Hence, $\mathbf{rename(r)}(I, O, S, F) = [\mathbf{r}(I), \mathbf{r}(O), \mathbf{r}(S), \mathbf{r}(F)]$.

We define $\mathbf{hide}(D)(I, O, S, F)$ to be $(I, O - D, \mathbf{del}(D)(S), \mathbf{del}(D)(F))$. Note that even if $S \cap F = \emptyset$, hiding some outputs may cause them to "overlap" (formally, it may be that $\mathbf{del}(D)(S) \cap \mathbf{del}(D)(F) \neq \emptyset$). For example, if $ab \in S$ and $ba \in F$, $a \in \mathbf{del}(\{b\})(S) \cap \mathbf{del}(\{b\})(F)$. In practical terms, this models a situation where both success and failure are possible, depending on unknown factors (or chance).

We say that two circuits are *behaviorally equivalent* when they have the same trace structures. It follows from the results of the previous chapter that for structural equivalence to imply behavioral equivalence (as it obviously should) it is necessary and sufficient that the operations on trace structures obey the laws of circuit algebra. The remainder of this section is devoted to proving the following theorem:

**Theorem 3.1.** *Prefix-closed trace structures are a circuit algebra.*

The following series of lemmas show prefix-closed trace structures are closed under the operations.

**Lemma 3.1.** *Prefix-closed trace structures are closed under hiding.*

**proof.** Let $T$ be any prefix-closed trace structure and $D$ be any subset of $O$, and let $T' = \mathbf{hide}(D)(T)$. $\mathbf{del}(D)$ preserves the regularity of $S'$ and $F'$. It is an immediate consequence of identity 3.6 that $S'$ and $P'$ are prefix-closed, as well. $P'$ is non-empty, since $\epsilon \in P$ and $\mathbf{del}(D)(\epsilon) = \epsilon$. Receptiveness is preserved by hiding, since none of the input symbols may be deleted ($I = I'$); Since $PI \subseteq P$, it is easy to see from the definition of $\mathbf{del}$ that $\mathbf{del}(D)(PI) \subseteq \mathbf{del}(D)(P)$, so $P'I' \subseteq P'$, also.  □

**Lemma 3.2.** *Prefix-closed trace structures are closed under inverse deletion.*

**proof.** Let $T$ be any prefix-closed trace structure, $D$ any finite subset of *Wires* disjoint from $A$, and let $T' = \mathbf{del}(D)^{-1}(T)$. As shown in Section 3.2, inverse deletion preserves regularity, so $S'$ and $F'$ are regular. It follows immediately from identity 3.7, above, that $P'$ and $S'$ are prefix-closed. $P'$ is also non-empty, because $\epsilon \in \mathbf{del}(D)^{-1}(\epsilon)$. Inverse deletion preserves receptiveness: Let $x'$ be any trace in $P'$. By definition, there exists an $x \in P$ such that $x = \mathbf{del}(D)(x')$. For every $a' \in I'$, either $a' \in I$ or $a' \in D$. If $a' \in I$ then $\mathbf{del}(D)(x'a') = xa'$ (since $a' \notin D$) and $xa' \in P$ (since $T$ is receptive), so $x'a' \in P'$. Otherwise, $a' \in D$, so $\mathbf{del}(D)(x'a') = x \in P$, so $x'a' \in P'$. Hence, in either case, $T'$ is receptive.  □

**Lemma 3.3.** *Prefix-closed trace structures are closed under intersection.*

**proof.** Let $T$ and $T'$ be any prefix-closed trace structures with appropriate alphabets, and let $T'' = T \cap T'$. Regular sets are closed under intersection, so $S''$ and $F''$ are regular. It follows immediately from inclusion property 3.8 that $P''$ and $S''$ are prefix-closed. To see that $T''$ is receptive, let $x$ be any member of $P'' = P \cap P'$, and let $a$ be any member of $I'' = I \cap I'$. Both $P$ and $P'$ must include $xa$, since both $T$ and $T'$ are receptive. So $P''I'' \subseteq P''$.  □

**Corollary.** *Prefix-closed trace structures are closed under composition.*

**Lemma 3.4.** *Prefix-closed trace structures are closed under renaming.*

**proof.** Obvious.  □

The remainder of this section shows that prefix-closed trace structures obey the laws of circuit algebra. Associativity and commutativity of composition (C1 and C2) follow immediately from the properties of $\cap$. C3 is easily proved by induction on sequences. The following lemma is helpful in proving C4 and C9:

**Lemma 3.5.** *For any sequence $x$ in $(A \cup D)^*$ and bijection $\mathbf{r}$ with domain $A \cup D$ that is naturally extended to sequences,* $\mathbf{r}[\mathbf{del}(D)(x)] = \mathbf{del}[\mathbf{r}(D)][\mathbf{r}(x)]$.

**proof.** By induction on sequences. Clearly, this holds when $x = \epsilon$. For the induction step, suppose that $\mathbf{r}[\mathbf{del}(D)(x)] = \mathbf{del}[\mathbf{r}(D)][\mathbf{r}(x)]$; since $a \in D$ iff $r(a) \in r(D)$, $\mathbf{r}[\mathbf{del}(D)(a)] = \mathbf{del}[\mathbf{r}(D)][\mathbf{r}(a)]$. $\mathbf{del}(D)$ and $\mathbf{r}$ both distribute over concatenation of sequences, so $\mathbf{r}[\mathbf{del}(D)(ax)] = \mathbf{del}[\mathbf{r}(D)][\mathbf{r}(ax)]$. $\square$

The next two lemmas show that C4 holds.

**Lemma 3.6.** $\mathbf{rename}(\mathbf{r})[\mathbf{del}(D)^{-1}(T)] = \mathbf{del}[\mathbf{r}(D)]^{-1}[\mathbf{rename}(\mathbf{r}|_{A \to \mathbf{r}(A)})(T)]$.

**proof.** Let $T_L$ and $T_R$ be the left- and right-hand sides of the equality to be proved. To see that $S_L = S_R$, let $x$ be any member of $S_L$. This is the case iff $\mathbf{del}(D)(x') \in S$, where $x' = \mathbf{r}^{-1}(x)$. This is true, in turn, iff $\mathbf{r}[\mathbf{del}(D)(x')] \in \mathbf{r}(S)$. By the previous lemma, this is true exactly when $\mathbf{del}[\mathbf{r}(D)][\mathbf{r}(x')] = \mathbf{del}[\mathbf{r}(D)](x) \in \mathbf{r}(S)$, which is true iff $x \in \mathbf{del}[\mathbf{r}(D)]^{-1}[\mathbf{r}(S)]$. Because $S \subseteq A^*$, this is equivalent to $x \in S_R$, so $S_L = S_R$.

Similarly, $F_L = F_R$. $\square$

**Lemma 3.7.** $\mathbf{rename}(\mathbf{r})(T \cap T') = \mathbf{rename}(\mathbf{r})(T) \cap \mathbf{rename}(\mathbf{r})(T')$.

**proof.** $\mathbf{r}$ is a bijection, so $\mathbf{r}(S \cap S') = \mathbf{r}(S) \cap \mathbf{r}(S')$ and $\mathbf{r}(F \cap F') = \mathbf{r}(F) \cap \mathbf{r}(F')$ $\square$

**Corollary.** *Prefix-closed trace structures obey C4.*

C5 is obvious from the definition of **rename**. C6 is easily proved by induction on sequences, using the recursive definition of **del**. C7 is obvious from the definition of **del**. C8, however, requires several lemmas.

**Lemma 3.8.** *If $E$, $D$, and $A$ are disjoint subsets of Wires, then*

$$\mathbf{del}(E)^{-1}[\mathbf{hide}(D)(T)] = \mathbf{hide}(D)[\mathbf{del}(E)^{-1}(T)].$$

**proof.** This is an immediate consequence of property 3.4. $\square$

**Lemma 3.9.** *If $X \subseteq A^*$ and $X' \subseteq (A \cup D')^*$ where $D' \cap A = \emptyset$, then $X \cap \mathbf{del}(D')(X') = \mathbf{del}(D')[\mathbf{del}(D')^{-1}(X) \cap X']$.*

**proof.** By the properties of inverse image functions, $X = \mathbf{del}(D')[\mathbf{del}(D')^{-1}(X)]$. From this and property 3.5, we can conclude that $\mathbf{del}(D')[\mathbf{del}(D')^{-1}(X) \cap X']$ is a subset of $X \cap \mathbf{del}(D')(X')$. To see inclusion in the other direction, suppose $x$ is any sequence in $X \cap \mathbf{del}(D')(X')$. Then $x \in X$ and there is a $y \in X'$ such that $x = \mathbf{del}(D')(y)$. But then $y \in \mathbf{del}(D')^{-1}(X)$, so $x \in \mathbf{del}(D')[\mathbf{dcl}(D')^{-1}(X) \cap X']$. $\square$

**Lemma 3.10.** *If $A \cap D' = A' \cap D = \emptyset$, then* $\mathbf{hide}(D)(T) \cap \mathbf{hide}(D')(T') = \mathbf{hide}(D \cup D')(T \parallel T')$.

**proof.** Let $T_L$ and $T_R$ be the left- and right-hand sides of the equation to be proved. First, we show that $S_L = S_R$. $S_L = \mathbf{del}(D)(S) \cap \mathbf{del}(D')(S')$ and, since $A' - A = D'$ and $A - A' = D$, $S_R = \mathbf{del}(D \cup D')[\mathbf{del}(D')^{-1}(S) \cap \mathbf{del}(D)^{-1}(S')]$. By the previous lemma, $\mathbf{del}(D)(S) \cap \mathbf{del}(D')(S') = \mathbf{del}(D)(S \cap \mathbf{del}(D)^{-1}[\mathbf{del}(D')(S')])$; by property 3.4, this is equivalent to $\mathbf{del}(D)(S \cap \mathbf{del}(D')[\mathbf{del}(D)^{-1}(S')])$ (note that $D \cap D' = \emptyset$). Applying the previous lemma a second time to the inner expression, we find this to be equivalent to $\mathbf{del}(D)(\mathbf{del}(D')[\mathbf{del}(D')^{-1}(S) \cap \mathbf{del}(D)^{-1}(S')])$ which is equal to $S_R$ by identity 3.3.

To see that $F_L = F_R$, let

$$X = \mathbf{del}(D)(F) \cap \mathbf{del}(D')(P'),$$
$$X' = \mathbf{del}(D)(P) \cap \mathbf{del}(D')(F'),$$
$$Y = \mathbf{del}(D')^{-1}(F) \cap \mathbf{del}(D)^{-1}(P'), \text{ and}$$
$$Y' = \mathbf{del}(D')^{-1}(P) \cap \mathbf{del}(D)^{-1}(F'),$$

so $F_L = X \cup X'$ and $F_R = \mathbf{del}(D \cup D')(Y \cup Y')$. Then by reasoning similar to above, $X = \mathbf{del}(D \cup D')(Y)$ and $X' = \mathbf{del}(D \cup D')(Y')$. Hence $X \cup X' = \mathbf{del}(D \cup D')(Y \cup Y') = F_R$, because deletion distributes over union, so $F_L = F_R$.  □

**Lemma 3.11.** *Prefix-closed trace structures obey C8.*

**proof.** Let $T'' = \mathbf{hide}(D)(T) \parallel \mathbf{hide}(D')(T')$, where $D' \cap A = D \cap A' = \emptyset$. By definition, $T''$ is equal to

$$\mathbf{del}[A' - (A - D)]^{-1}[\mathbf{del}(D)(T)] \cap \mathbf{del}[A - (A' - D')]^{-1}[\mathbf{del}(D')(T')].$$

Since $A' \cap D = \emptyset$, $A' - (A - D) = A' - A$, so by Lemma 3.8, this is equal to

$$\mathbf{del}(D)[\mathbf{del}(A' - A)^{-1}(T)] \cap \mathbf{del}(D')[\mathbf{del}(A - A')^{-1}(T')].$$

Hence, by the previous lemma and the definition of **compose**, $T'' = \mathbf{del}(D \cup D')(T \parallel T')$.
□

Finally, C9 is easily proved by induction on sequences.

## 3.4.  Examples

In this section we present several examples to illustrate the use of prefix-closed trace structures. However, before doing so, there is a general issue that needs to be addressed. In all interesting cases, the trace structure of a device depends on its initial state (the logical values on its inputs and outputs, and possibly unobservable internal state). We assume that the circuits have hidden initialization circuitry which reliably leads to a well-defined initial state (or set of initial states), and that circuit descriptions include a description of the initial conditions.

### 3.4.1. Boolean Gates

Consider an NOR gate with inputs $a$ and $b$ and an output $c$. If $a = b = 0$, $c = 1$, and there is a transition on $a$, $c$ (the output) will remain high for some period of time. If $a$ stays high, $c$ will eventually go low. During this time, $b$ can change several times without altering the outcome, since the output is completely determined by the value of $a$. However, if the circuit is in the state $a = c = 1$, $b = 0$, and if there is a second transition on $a$, the circuit could exhibit a *hazard*: the output voltage could fall part of the way to $c = 0$ and then reverse direction, or it could go to 0 for an arbitrarily brief time and then return to 1. Hazards can cause severe problems in asynchronous circuits; in some cases, they can render digital models invalid.

A formal model of gate behavior should classify hazardous input transitions as failures. A gate has a set of inputs, $I$, and a single output. It can be thought of as an element that instantaneously computes the value of a Boolean function and then sets the output to this value after an arbitrary delay. A gate is *stable* if the value of the Boolean function is the same as the current output of the gate; otherwise, it is *unstable*. If a gate is stable, the output cannot change unless the input changes. If a gate is unstable, the output can change. After the output changes, the gate will be stable again (until the next input change). An input change that makes the gate stable *before* the output change occurs should be a failure (it may cause a hazard). However, an input may change successfully if the gate is unstable before and after the input change.

Formally, the logical values of the wires at any time can be represented as a function $b$ that maps each wire to a logical value: $b: A \rightarrow \{0, 1\}$. This function is called a *wire valuation*. The set of all possible wire valuations is $\mathbf{B} = [A \rightarrow \{0, 1\}]$. Each gate is characterized by a Boolean function (call it $f$), which says what the output of the gate will eventually be, given a particular set of logical values for the inputs. Since this Boolean function depends only on the input values, but the functions in $\mathbf{B}$ give values for the input and the output, the wire valuations in the domain of $f$ are restricted to the inputs of the gate, $I$. Formally, the type of $f$ is

$$\{\mathbf{b}|_I \mid \mathbf{b} \in \mathbf{B}\} \rightarrow \{0, 1\}.$$

The trace structure for a gate depends on the initial logical values of the inputs and the output (call the valuation $\mathbf{b}_0$). The most direct way to define the trace structure is to describe the finite state automata that accept its trace sets. Let the trace structure be $T$. We define a single state graph, $\mathcal{M}$, which accepts either $S$ or $F$ depending on the choice of final states.

The alphabet $A$ of $\mathcal{M}$ is the same as that of $T$ (which is also called $A$). The set $Q$ of states of $\mathcal{M}$ is $\mathbf{B} \cup \{q_{\text{fail}}\}$, where $q_{\text{fail}}$ is a special state not in $\mathbf{B}$. A trace whose run ends in a $\mathbf{B}$ state is successful, and a trace ending in $q_{\text{fail}}$ is a failure. The initial state $q_0$ is $\mathbf{b}_0$.

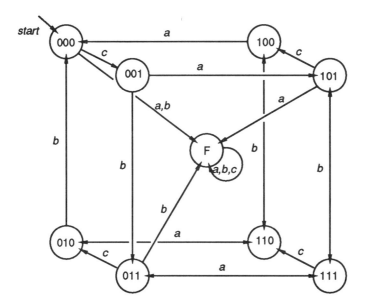

**Figure 3.1.** NOR Gate and Automata

Trace sets are sequences of transitions, not logical values. However, given a logical valuation and a transition, it is easy to determine the logical valuation after the transition — the value of the wire on which the transition occurred is complemented. We define a function that does this. **bitcomp**: $\mathbf{B} \times A \rightarrow \mathbf{B}$ is defined so that $\mathbf{bitcomp}(\mathbf{b}, a)(a) = \overline{\mathbf{b}(a)}$ (the Boolean complement of $\mathbf{b}(a)$) and $\mathbf{bitcomp}(\mathbf{b}, a)(b) = \mathbf{b}(b)$ if $b \neq a$.

Call the output of the gate $c$ ($O = \{c\}$). A state $\mathbf{b}$ is *stable* if the value of its Boolean function is equal to the value of $c$: $f(\mathbf{b}|_I) = \mathbf{b}(c)$. The next-state function $\mathbf{n}$ of $\mathcal{M}$ is defined to allow any input transition from a stable state: if $\mathbf{b}$ is stable and $a \in I$, $\mathbf{n}(\mathbf{b}, a) = \mathbf{bitcomp}(\mathbf{b}, a)$; an output transition is impossible in a stable state, so if $\mathbf{b}$ is stable and $a \in O$, $\mathbf{n}(\mathbf{b}, a)$ is not defined. If a state is unstable and an input transition would take it to a stable state, the transition should go to the failure state, instead: if $\mathbf{b}$ is unstable and $\mathbf{bitcomp}(\mathbf{b}, a)$ is stable, $\mathbf{n}(\mathbf{b}, a) = q_{\text{fail}}$. Otherwise, $\mathbf{n}(\mathbf{b}, a) = \mathbf{bitcomp}(\mathbf{b}, a)$ (including when $a \in O$, because an output is possible in an unstable state). Finally, $\mathbf{n}(q_{\text{fail}}, a) = q_{\text{fail}}$ for all $a \in A$ — if a trace has a failure prefix, it is a failure.

The set of traces accepted by $\mathcal{M}$ depends on its set final states, $Q_F$. To make $\mathcal{M}$ accept the set of successful traces, $S$, set $Q_F$ to $\mathbf{B}$ (so $q_{\text{fail}}$ is the only state that is not final). Since the only successor state of $q_{\text{fail}}$ is $q_{\text{fail}}$ itself, $S$ is prefix-closed. To make $\mathcal{M}$ accept the set of failures, define $Q_F = \{q_{\text{fail}}\}$. The set of failures are those traces where an input takes the gate from an unstable state to a stable one — in other words, the illegal transitions. An automaton for $P$ can be had by making *all* of the states in $Q$ final. $P$ is also prefix-closed. It is easy to see that in the resulting machine $\mathbf{n}(\mathbf{b}, a)$ is defined for every $\mathbf{b} \in \mathbf{B}$ $a \in I$, so $P$ is receptive, too.

A state diagram for a NOR gate appears in Figure 3.1. The inputs are $a$ and $b$ and the output is $c$. The successful states are labeled with the values of $a$, $b$, and $c$, in that order. The initial state is 000, and $q_{\text{fail}}$ is labelled with $F$.

### 3.4.2. Ring Oscillator

Figure 3.2a depicts a *ring oscillator*. It can be constructed by composing an inverter $T_I$ that has input $a$ and output $b$ with a non-inverting buffer $T_B$ that has input $b$ and output $a$.

If $a = b = 0$ initially in each case, the inverter, buffer, and their composition, the ring oscillator, have the same success sets. The state diagram in Figure 3.2b accepts this set if all four states are made final. The inverter, buffer, and ring oscillator have different failure sets, however. To accept $F_I$, states 00 and 11 would have transitions to a failure state on $a$. To accept $F_B$, failing transitions on $b$ should be added to 01 and 10. Since $A_I = A_R$, $T_I \parallel T_R = T_I \cap T_R$. The trace structure for the composite circuit has $I = \emptyset$, $O = \{a, b\}$, $S = \mathbf{pref}[(ba)^*]$, and $F = \emptyset$. Note that there are no inputs or failure

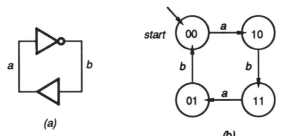

(a)

(b)

**Figure 3.2.**   Ring Oscillator

traces, and that the ring oscillator generates an endless sequence of alternating $b$ and $a$ transitions. This faithfully represents the behavior of physical ring oscillators.

If the $b$ wire is hidden, the resulting structure is $(\emptyset, \{a\}, [a^*], \emptyset)$, representing an endless stream of $a$ transitions.

### 3.4.3. Set-Reset Latch

Figure 3.3a is a circuit diagram of a set-reset latch, implemented as a cross-wired NOR. The circuit should meet the following *informal* specification (a formal specification appears in the next chapter): the circuit has two modes*: 'set', in which $c = 1$ and $d = 0$, and 'reset', in which $c = 0$ and $d = 1$. If the circuit is in either of these modes, and both inputs are 0, the circuit stays in that mode. If it is in the 'reset' mode and the $a$ input rises, it should eventually change to the 'set' mode; $a$ can return to 0 only after the circuit is set. Similarly, $b$ resets the latch if it is set, and is not allowed to return to zero until it is reset. The desired behavior is undefined if both inputs are 1, if both outputs are 1, or if an input which would change the state of the latch returns to 0 before the circuit settles into the appropriate mode.

The trace structures for the *actual* behavior of the two gates can be obtained by composing the trace structures of two individual NOR gates. Note that the initial state for gate G1 is 001 and gate G2 is 010. These two trace structures must be renamed in the obvious way and then composed. In the result, $I = \{a, b\}$ and $O = \{c, d\}$. The resulting trace sets ($S$, $F$, and $P$) are described by the automaton depicted in Figure 3.3b. The successful states are labeled with the values for $a$, $b$, $c$, and $d$, in that order.

The initial state is 0001, in which the latch is reset and both inputs are 0. Failing transitions on input symbols are not shown; however, failing transitions on *output* transitions

---

*The word "states" has already been used in the definition of finite automata.

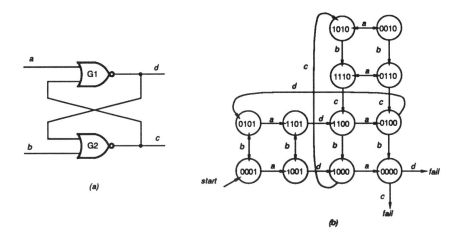

**Figure 3.3.**   Set-Reset Latch Implementation

are drawn as arrows leading to *"fail"*. This part of the machine is not shown because it is both difficult to draw and irrelevant to the discussion below.

As a typical example of a trace in the composite trace structure consider the state 1001, in which G1 is unstable and G2 is stable. From this state, a *d* output is possible (and successful) because of the instability of G1, a *b* transition is successful because G2 is stable, but an *a* transition is a failure because it would otherwise take G1 from an unstable state to the stable state 0001.

The resulting behavior appears to meet our informal specification of a latch, but is much more detailed. Specific failure modes appear in Figure 3.3b in cases with which the informal specification is not concerned. Also, perhaps the implementation could produce useful results in situations not required by the specification. For example, the trace *adcba* is not allowed by the specifications, but in the implementation has the same result as *adcab*, with no possibility of failure along the way. Thus, the latch could be used in a circuit in which *a* and *b* are (sometimes) concurrent, an application that was not foreseen in the specifications.

To see an example of a trace that is both a success and a failure, consider the result of hiding the output *c*. The trace *adcb* is a success, but *adb* is a failure. Hence, in the new circuit, $adb = \mathbf{del}(\{c\})(adcb) = \mathbf{del}(\{c\})(adb)$ is both a success and a failure. There is a meaningful practical interpretation of this result. If *c* is hidden and if an input *b* occurs after *ad*, it is impossible to tell if the restriction "*b* does not change after *a* until *d* and *c* have changed" is met. If the sender is lucky, the *b* will arrive after the invisible *c* transition and all will be well — but it is also possible that the *b* will arrive before *c*, causing a failure. In essence, the circuit appears to choose nondeterministically between success and failure, so both possibilities are represented in the trace structure.

A classic anomaly in circuits of this kind is the *metastable state* [23]. In general, metastable states can occur when a circuit is expected to choose between mutually-exclusive states. An effective way to put this circuit in a metastable state is to raise one of the inputs to flip the state (say, from reset to set), but remove the input while the circuit is in the unstable $c = 0$, $d = 0$ state. If this occurs when the circuit is in the reset or set mode (0001 or 0010) this input takes it to state 0000. The circuit is neither set nor reset and cannot decide between them based on its inputs, because everything is completely symmetric; furthermore, if it decides to raise either *c* or *d*, it goes into a state where both gates are stable, disabling the other output (so the choice is mutually-exclusive). In this sort of dilemma, the circuit may take an arbitrarily long time to decide between the two states, but worse, it may "split the difference" by outputting voltages between the thresholds for logical 0 or 1 or oscillating between them. This sort of behavior is disastrous in an asynchronous circuit.

The phrase "disabling the other output" is the key to preventing metastable states in circuits of Boolean gates. To disable a pending gate output, the inputs to the gate must change from an unstable to a stable state before the output occurs — but this a failure in the trace structures for gates and consequently in any composition containing them. In the diagram there are two output transitions from 0000; both lead to the failure state.

It is not always possible in practice to avoid metastable states. However, it *is* possible by using analog circuit techniques to design components which handle metastable states gracefully — so that outputs behave consistently with a digital model. (However, arbitrary delays in resolving the metastable state cannot be avoided.) Such devices are called *mutual-exclusion circuits* or *arbiters*; there are many different types. Digitally well-behaved circuits can be designed if an arbiter is used whenever a choice is to be made between two mutually-exclusive states.

As we shall see in the next example, it is possible to describe an arbiter with trace structures, but it is impossible to construct this behavior by composition of gate behaviors. The result is that metastable states in a composite circuit are flagged as failures unless they are handled by special devices.

### 3.4.4. Mutual-Exclusion Element

A mutual-exclusion element is a type of arbiter which raises at most one of two outputs in response to transitions on two inputs. If the inputs are independent and if they can arrive at any time, such a device has an inherent metastable state. The element must be a basic circuit, since the implementation requires analog circuit design techniques [69,70] which cannot be modeled using trace theory (or any other theory based on logical values of signals).

Figure 3.4a shows a mutual-exclusion element, with inputs $r1$ and $r2$ and outputs $a1$ and $a2$. If all signals are initially low, and $r1$ but not $r2$ is raised, the element should raise $a1$; similarly, if $r2$ but not $r1$ is raised it should raise $a2$. If *both* inputs go high, the element nondeterministically chooses *exactly one* of the two outputs to raise. It is never the case that both outputs are high. If one of the outputs returns to zero and the other input is high, the element should raise the output for the other input. Each input must remain high until the appropriate output goes high, at which point the input can return to zero. The corresponding output should then return to zero. The input cannot go high again until the corresponding output returns to zero.

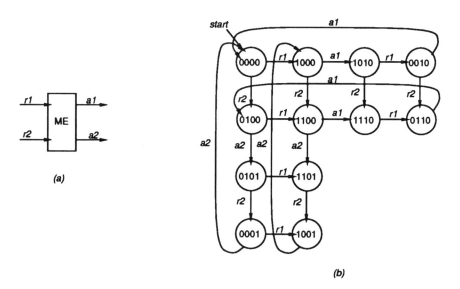

(a)

(b)

**Figure 3.4.**   Mutual-Exclusion Element

A state diagram for a mutual-exclusion element is shown in Figure 3.4b. As usual, the failures on input transitions are not shown. The mutually exclusive choice between outputs occurs at state 1100 (requests from both inputs are outstanding, but neither output has been raised). Note that none of the four states having $c = d = 1$ appear in the state graph.

### 3.4.5. C Element

The *C element* is an important primitive in speed-independent circuit design (Figure 3.5a). This circuit holds its output until both inputs have the logical value that is the complement of the output; it then changes the output to be the same as the inputs. C elements are generally used when some part of a circuit wants to wait for two concurrent computations to complete. In each case, completion is signaled by a transition on some wire, which is connected to the inputs of the C element. The output of the C element changes when transitions have occurred on both inputs.

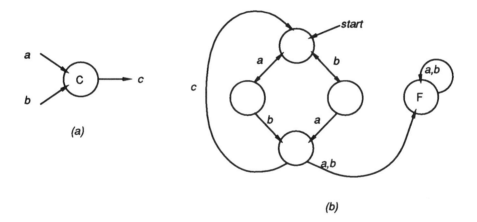

*(a)*

*(b)*

**Figure 3.5.**   C Element

An automaton is shown in Figure 3.5b. The failures are those cases where the output of the C element has been enabled (both inputs have gone to the opposite value of the output), but an input changes (again) before the output changes. This is a potential hazard, because the C element may change the output part of the way before the input change disables it.

### 3.4.6. Vending Machine

This example is not a practical circuit, but answers an objection that may occur to readers familiar with CCS. We consider two "vending machines": one inputs money ($a$) and lets the customer select one of two items by taking inputs $b$ and $c$ (and for simplicity in the example, never works again); the other takes money, then makes an internal decision about which product it will allow the customer to choose (without telling him). The customer selects item $b$ or $c$, and either gets it or not, depending on the decision made by the machine.

These two behaviors can be represented using the CCS-like trees of Figures 3.6a and 3.6b (these are *not* automata). Figure 3.6a accepts an $a$ input and then either $b$ or a $c$, depending on what the environment chooses. In Figure 3.6b, when the machine accepts the $a$, it nondeterministically chooses whether to respond to a $b$ or $c$ input. It seems that these behaviors should be distinguished, and indeed they are distinguished by CCS's concept of observational equivalence. However, the traces ($ab$ and $ac$) of the two trees are identical. Does this mean that the trace representation of the behaviors is insufficient?

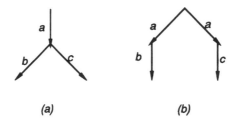

(a)                              (b)

**Figure 3.6.** "Vending Machines"

Our notion of behavioral equivalence distinguishes the two trees, also, because equivalence depends on *both* the success and failure sets. The *successful* traces of the two trees are the same. However, the failure sets are not. In Figure 3.6a, the failure traces are

$$[aa + b + c + (ab + ac)(a + b + c)](a + b + c)^*.$$

In Figure 3.6b, the failures include the failures for the other tree, and also $ab$ and $ac$ ($ab$ is a failure if the vending machine chooses the right branch, and $ac$ is a failure if it chooses the left branch). So, the behaviors of the two machines have different trace structures. In 3.6b, $ab$ may be a success or a failure; the customer who inputs a $b$ has no control over the result.

# Chapter 4

# Verification

## 4.1.  Introduction

To verify a circuit is to demonstrate that the actual behavior satisfies some specification. Chapter 3 has solved the problem of describing actual behaviors. This chapter shows how to specify the desired behavior and how to verify automatically whether the actual behavior is consistent with the specification.

Instead of inventing a second formalism for specifications, we use the one we already have: trace structures. A specification is regarded as an idealized component, which can can be used (perhaps conceptually) in a design. Another trace structure represents an implementation of the specification if it can be *substituted* for the idealized component while preserving the important properties of the larger design. Using the idea of safe substitution, the same formalism can be employed both for semantics and specification.

The concept of safe substitution makes hierarchical verification very easy and natural. A specification of a circuit at one level of abstraction can be a description of a primitive at the next higher level. For example, consider a flip-flop implemented as a collection of gates and used in a larger circuit. The verification of the larger circuit can be split into a low-level phase and a high-level phase. At the low level, it is verified that the gates properly implement the flip-flop specification. At the high level, it is verified that the flip-flop specification and other components properly implement the larger circuit. At this level, the gate implementation of the flip-flop is hidden — only the flip-flop specification is used. Verifying the circuit in this way has important advantages. If the flip-flop is used more than once or in several circuits, the low-level verification does not need to be repeated. Moreover, if a different implementation of the flip-flop is adopted, the designs that use flip-flops as primitives do not need to be re-verified.

Trace structure specifications are highly compositional. Two implementations can be shown to satisfy less detailed interface specifications; it is then immediately true that the composition of the implementations satisfies the composition of the specifications — this does not need to be checked separately. Also, if symbols are hidden in the implementation, the result is sure to satisfy the specification with the same symbols hidden (likewise with renaming).

Also, trace structures can have restrictions on their allowed inputs. These restrictions are assumptions about the environments with which the circuit will be composed. Consequently, specifications are relative to assumptions about the environment. This is a very useful property: known constraints on the environments in which a circuit will be used may increase the range of implementation alternatives.environmental assumptions

In the following sections, the second formally defines safe substitution. A relation called *conformation* is defined which holds when one trace structure can be substituted safely for another. In this situation, the first structure is said to *conform to* the second.

Sometimes two different trace structures cannot be distinguished by conformation. This defines an equivalence relation between trace structures called *conformation equivalence*. In the fourth section, two transformations are described which can be used to simplify a trace structure while maintaining conformation equivalence to the original. The fourth section also explains how a trace structure represents assumptions about its environment, and how to characterize the most general environment that satisfies that assumption. This is a very important result because it leads to a decision procedure for conformation. It is also a powerful tool for proving other properties of trace structures, for example that the simplifications yield a canonical form for conformation equivalence. The fifth section gives examples to illustrate conformation.

A specification is *delay-insensitive* if it does not change when delays are attached to its inputs and outputs. Delay insensitivity has received a great deal of attention in previous work on trace theory. In the sixth section, a simple definition of delay insensitivity is given, using the results of the previous sections. Delay insensitivity of a trace structure can be checked automatically.

The seventh (and final) section shows that the set of all trace structures with particular inputs and outputs is a distributive lattice. The lattice operations of *meet* and *join* provide additional interesting ways to combine trace structures.

## 4.2.    Conformation

To formalize the definition of safe substitution, we first define an *expression context*: it is a circuit algebra expression with a free variable (the variable is written $\alpha$). When a trace structure is substituted for the free variable, the expression can be evaluated to give a trace structure for the expression. $\alpha$ has inputs $I_\alpha$ and outputs $O_\alpha$; any trace structure $T$ that is substituted for $\alpha$ must have $I = I_\alpha$ and $O = O_\alpha$. Because of this, it is possible to determine whether the alphabets of its subexpressions meet the requirements of the operations and to find the inputs and outputs of the entire context. An example of an expression context is $\mathbf{hide}(\{a\})(T \parallel \alpha)$. A generic expression context is written as $\mathcal{E}[\,]$, and the expression resulting when $T$ is substituted for $\alpha$ is written $\mathcal{E}[T]$.

What is a "safe" substitution? We say that a trace structure is *failure-free* when its failure set is empty. A safe substitution is a substitution that preserves failure-freedom. Formally, a trace structure $T$ *conforms to* $T'$ $(T \preceq T')$ if $I = I'$, $O = O'$, and for *all* expression contexts $\mathcal{E}[\,]$, if $\mathcal{E}[T']$ is failure-free, so is $\mathcal{E}[T]$. Intuitively, if a system using $T'$ cannot fail, neither can a system using $T$. The relation is called *conformation*. (We use the term "conforms to" instead of "meets" or "satisfies" to avoid confusion the terminology of logic and lattice theory.)

Only a small subclass of context expressions actually needs to be considered to prove conformation. The context expression can be reduced to a composition with a single trace structure, called an *environment*, which has inputs and outputs which are complementary to the free variable: $T \preceq T'$ if and only if for every $T''$ having $I'' = O$ and $O'' = I$, if $T' \cap T''$ is failure-free, so is $T \cap T''$.

The rest of this section proves these results. It is useful to define another equivalence relation between trace structures: $T$ and $T'$ are *FF-equivalent* if and only if they are either both failure-free or both not failure free. Unlike behavioral equivalence, FF-equivalence does not imply matching alphabets. Obviously, behavioral equivalence implies FF-equivalence.

Two *contexts* $\mathcal{E}[\,]$ and $\mathcal{E}'[\,]$ are *behaviorally equivalent* if for every trace structure $T$ we have $\mathcal{E}[T] = \mathcal{E}'[T]$. $\mathcal{E}[\,]$ and $\mathcal{E}'[\,]$ are *FF-equivalent* if for every trace structure $T$, $\mathcal{E}[T]$ is FF-equivalent to $\mathcal{E}'[T]$. In either case, the free variables of the two contexts must have the same inputs and outputs.

**Lemma 4.1.** *If $\mathcal{E}[\,]$ is an expression context, it is FF-equivalent to the expression contexts* $\mathbf{hide}(D)(\mathcal{E}[\,])$, $\mathbf{rename}(\mathcal{E}[\,])$, *and* $\mathbf{del}(D)^{-1}(\mathcal{E}[\,])$, *when they are defined.*

**proof.**  Obvious.    □

**Lemma 4.2.** *$T \preceq T'$ iff $I = I'$, $O = O'$, and for every $T''$ such that $I = O''$ and $O = I''$, if $T' \cap T''$ is failure-free, so is $T \cap T''$.*

**proof.** Let $\mathcal{E}[\ ]$ be any expression context. By Lemma 2.6, every expression can be reduced to a structurally equivalent (and consequently FF-equivalent) normal-form expression using the laws of circuit algebra. Trace structures are a circuit algebra, so there is an FF-equivalent expression $\mathcal{E}_1[\ ]$ in normal form.

Let $\mathcal{E}_2[\ ]$ be the subexpression of $\mathcal{E}_1[\ ]$ such that $\mathcal{E}_1[\ ] = \textbf{hide}(D)(\mathcal{E}_2[\ ])$; then by Lemma 4.1, $\mathcal{E}_1$ and $\mathcal{E}_2$ are FF-equivalent. $\mathcal{E}_2[\ ]$ consists of a composition of renamed trace structures. **compose** is associative and commutative, so the $n$-way composition of $\mathcal{E}_2[\ ]$ can be grouped into two parts: a **rename** expression with the free variable, and the others. The expression for the others can be evaluated to yield a single trace structure $T_1$, so there is an expression $\mathcal{E}_3[\ ]$ of the form $\textbf{rename}(\textbf{r})(\alpha) \,\|\, T_1$ which is behaviorally equivalent to $\mathcal{E}_2[\ ]$. If we let $T_2 = \textbf{rename}(\textbf{r}^{-1})(T_1)$, and $\mathcal{E}_4[\ ] = \alpha \,\|\, T_2$, $\mathcal{E}_3[\ ]$ is FF-equivalent to $E_4[\ ]$ by Lemma 4.1.

At this point, we have reduced the context to $\alpha \,\|\, T_2$, but the alphabet of $T_2$ does not fulfill the requirements of the desired context. However, we do know that $O_\alpha \cap O_2 = \emptyset$. The rest of the proof massages the alphabet of the trace structure with which $\alpha$ is composed.

Let $A_\alpha = I_\alpha \cup O_\alpha$. We find an FF-equivalent context $\mathcal{E}_5[\ ] = \alpha \,\|\, T_3$ such that $A_\alpha \subseteq A_3$. Set $T_3 = \textbf{del}(A_\alpha - A_2)^{-1}(T_2)$, so that $A_3 = A_\alpha \cup A_2$. By Lemma 4.1, $\mathcal{E}_5[\ ] = \alpha \,\|\, T_3$ is FF-equivalent to $\mathcal{E}_4[\ ]$.

Next, we find $\mathcal{E}_6[\ ] = \alpha \,\|\, T_4$ where $A_4 = A_3$ and $I_4 = O_\alpha$ (some of $T_3$'s inputs are converted to outputs). This is done by composing $T_3$ with a "generator" circuit that produces all possible outputs in $I_3 - O_\alpha$. Let $T_G$ be the trace structure having $O_G = I_3 - O_\alpha$, $I_G = A_3 - O_3$ (so $A_G = A_3$), $S = A_3^*$, and $F = \emptyset$. Let $T_4 = T_3 \cap T_G$. Clearly, $S_4 = S_3 \cap S_G = S_3$. Since $P_G = A_3^*$ and $F_G = \emptyset$, $F_4 = (F_3 \cap P_G) \cup (P_3 \cap F_G) = F_3$. Also, $P_4 = P_3 \cap P_G = P_3$. Any composition of the form $\alpha \,\|\, T_4$ has exactly the same $F$, $S$, and $P$ sets as $\alpha \,\|\, T_3$; only the alphabets are different. Hence $\mathcal{E}_6[\ ]$ is FF-equivalent to $\mathcal{E}_5[\ ]$.

We now have $O_\alpha = I_4$ and $I_\alpha \subseteq O_4$. All that remains to complete the lemma is to remove the extra outputs of $T_4$, which is easily done by setting $T''$ to $\textbf{hide}(O_4 - I_\alpha)(T_4)$. $\alpha \,\|\, T'' = \alpha \cap T''$ is FF-equivalent to $\mathcal{E}_6[\ ]$ by Lemma 4.1, completing the proof. $\quad\square$

If implementations are to be freely substituted for specifications, it is important that if $T \preceq T'$, then $\mathcal{E}[T] \preceq \mathcal{E}[T']$ (if $\mathcal{E}[T]$ is defined). This is shown in the next theorem.

**Theorem 4.1.** **compose**, **rename**, *and* **hide** *are monotonic with respect to conformation.*

This is proved in the next three lemmas.

**Lemma 4.3.** *If $T \preceq T'$ and $T''$ is any trace structure, then $T \parallel T'' \preceq T' \parallel T''$.*

**proof.** Let $T'''$ be any environment such that $(T' \parallel T'') \cap T''' = (T' \parallel T'') \parallel T'''$ is failure-free. Then, by associativity of **compose** (C1), $T' \parallel (T'' \parallel T''')$ is failure-free. Then, by the definition of conformation and by C1, $(T \parallel T'') \cap T'''$ is failure-free. Hence, by Lemma 4.2, $T \parallel T'' \preceq T' \parallel T''$.   □

**Lemma 4.4.** *If $T \preceq T'$, then* **hide**$(D)(T) \preceq$ **hide**$(D)(T')$.

**proof.** Let $T''$ be any environment such that **hide**$(D)(T') \parallel T''$ is failure-free. Since $A'' = A - D$, $A'' \cap D = \emptyset$, so by C7 and C8, **hide**$(D)(T') \parallel T'' = $ **hide**$(D)(T' \parallel T'')$. By Lemma 4.1, this is FF iff $T' \parallel T''$ is. But, since $T \preceq T'$, $T \parallel T''$ is failure-free, and so is **hide**$(D)(T \parallel T'')$. This is equal to **hide**$(D)(T) \parallel T''$, by C7 and C8, so **hide**$(D)(T) \preceq$ **hide**$(D)(T')$.   □

**Lemma 4.5.** *If $T \preceq T'$, then* **rename**$(r)(T) \preceq$ **rename**$(r)(T')$.

**proof.** Obvious.   □

## 4.3.  Conformation Equivalence and Simplifications

Our choice of symbols suggests that the $\preceq$ relation is a partial order. However, this is not the case; it is reflexive and transitive, as is clear from the use of implication in the definition. However, it is not antisymmetric — it is a *pre-order*. Hence, conformation induces a nontrivial equivalence relation on trace structures. $T$ is *conformation equivalent* to $T'$ ($T \sim_c T'$) when $T \preceq T'$ and $T' \preceq T$. $T \sim_c T'$ does not necessarily imply that $T' = T$. Intuitively, trace structures have more information in them than is really necessary to determine conformance. For example, one may be able to tell by looking at a trace structure that after it receives an illegal input, it produces no more than five output signals. But all we *need* to know to determine conformance is the exact circumstances under which the first failure occurs — the details of the circuit behavior after this failure are completely irrelevant. The topic of this section is how to remove irrelevant information from a trace structure.

There are two transformations to reduce a prefix-closed trace structure to a *canonical form* for conformation equivalence. Two trace structures in canonical form are equal exactly when they are conformation equivalent (this is proved in the next section).

The transformations are best introduced through an example. The example circuit is a pair of inverters in series, as in Figure 4.1a. The state graph for a single inverter, starting in a stable state, appears in Figure 4.1b. This automaton accepts the same trace sets as the automaton for an identity gate, as defined in Section 3.4 (it has been reduced from five states to three). An automaton for the composite circuit appears in Figure 4.1c. In the initial state, both inverters are stable (e.g. $a = 0$, $b = 1$, and $c = 0$). Every state in the composite has two labels: the names of the states in the original two automata. A trace reaches one of the states in the automaton if the trace is possible in both inverters. The successful traces in the composite are those ending in states labeled $q_i q'_j$ (where $i, j \in \{0, 1\}$); the failures are those ending in states labeled with $F$ or $F'$.

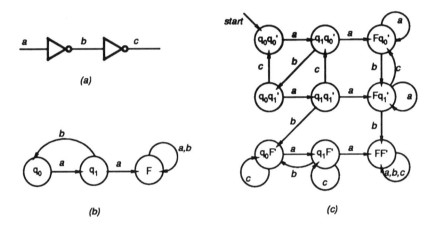

**Figure 4.1.**  Two-inverter Chain

The first and most interesting simplification is called *autofailure manifestation*. In Figure 4.1c, the trace *aba* reaches the state $(q_1, q'_1)$, which is successful. However, from this state the machine can output a *b*, which causes a failure (specifically, in the second inverter). Once this state has been reached, there is *no way* for the environment to prevent the *b* output and the failure. The trace *aba* is called an *autofailure* because the circuit causes its own failure. Autofailure manifestation adds all autofailures to the failure set — it makes them explicit. Autofailure manifestation also adds any trace that has a failure as a prefix to the failure set.

More precisely, an autofailure is any trace $x$ such that there is a trace $y \in O^*$ such that $xy \in F$. The set of all such traces is $F/O^*$ (note that since $\epsilon \in O^*$, this set includes $F$). All suffixes can be added to a set by concatenating with $A^*$. Hence, the simplification is to redefine the set of failures to be $(F/O^*)A^*$ (this set includes $F/O^*$). The inputs, outputs, and success set remain unchanged.

Unless $S = \emptyset$, every failure trace has a prefix that is a success. After autofailure manifestation, the symbol which converts the success to a failure is always an input — the environment has sent a transition to the circuit that the circuit could not handle. We call this *choking*. If $x$ is a success, $a$ is an input symbol, and $xa$ is a failure, we call $xa$ a *choke*.

The second and more obvious transformation is to make the success and failure sets of a trace structure disjoint. Recall that if $x \in S \cap F$ in a trace structure $T$, $x$ can be either a success or a failure; in essence, the circuit chooses nondeterministically between them. For verification, a circuit that "might fail" on a particular trace is just as bad as a circuit that always fails — in either case, if $x$ is allowed to appear in a particular composite circuit containing $T$, that composite is not failure-free.

The transformation is called *failure exclusion*; it sets the success set to $S - F$. Failure exclusion is only valid when the trace structure to which it is applied satisfies the additional property: $FA^* \subseteq F$; otherwise, there is a chance that the success set of the result will not be prefix-closed. One way to make sure this condition is satisfied is to do autofailure manifestation first. Note that neither the $F$ set nor the $P$ set changes.

Figure 4.2 shows the automaton for the two inverters of Figure 4.1 after all of the transformations have been applied. More than half of the states have been eliminated.

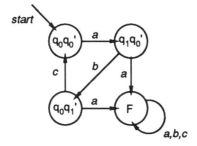

**Figure 4.2.** Two-inverter Chain After Simplification

The result of these transformations is a trace structure in canonical form, which can be characterized directly: a *canonical prefix-closed trace structure*, $T$, is a prefix-closed trace structure satisfying the additional requirements $F/O \subseteq F$, $FA \subseteq F$ and $S \cap F = \emptyset$. $F/O \subseteq F$ implies $(F/O)/O \subseteq F/O$ and so on, so the first condition is equivalent to $F/O^* \subseteq F$. Similarly, $FA \subseteq F$ if and only if $FA^* \subseteq F$.

A nice property of canonical prefix-closed trace structures is that the set of failures is completely determined by the set of successes. If $T$ is in canonical form, the set of failures is $[(SI \cup \{\epsilon\}) - S]A^*$. Intuitively, the set $(SI \cup \{\epsilon\}) - S$ is the set of *minimal failures* (members of $F$ that do not have failures as prefixes). If $S$ is empty, there is exactly one minimal failure: $\epsilon$. Otherwise, the minimal failures are exactly the *chokes*: $SI - S$.

Since the presence of the failure set does not contribute any information to a canonical trace structure, a reduced representation can be defined: the *simple prefix-closed trace structures*. A simple prefix-closed trace structure is a triple $T = (I, O, S)$. We define abbreviations $A = I \cup O$ (as before), $F = [(SI \cup \{\epsilon\}) - S]A^*$, and $P = S \cup F$. The conversion between simple and canonical trace structures works both ways: any simple trace structure can be converted to a canonical trace structure simply by making the implicit failure set an explicit fourth element of the structure. To avoid confusion, we sometimes call trace structures that are not necessarily canonical or simple *general* trace structures.

For most applications, it is not useful to distinguish that are conformation-equivalent trace structures. Hence, an interpretation of the operations of circuit algebra that always maintains trace structures in canonical form would be useful. To define such a thing, apply the operations as defined for general trace structures, then simplify the results to canonical form. The program described in Chapter 5 uses this representation. The proof that these definitions obey the laws of circuit algebra is deferred until the next section.

The rest of this section is devoted to proving these claims. The following extremely useful lemma gives a simple sufficient (but not necessary) condition for conformation between general prefix-closed trace structures. The condition $F \subseteq F'$ assures that if the environment does not cause a failure in $T'$, it will not cause a failure in $T$, either. The condition $P \subseteq P'$ assures that if $T'$ does not cause a failure in the environment, $T$ will not cause one, either.

**Lemma 4.6.**  $T \preceq T'$ if $I = I'$, $O = O'$, $F \subseteq F'$, and $P \subseteq P'$.

**proof.** Suppose the conditions are all met. Let $T''$ be any trace structure such that $I'' = O$ and $O'' = I$, and suppose $T' \cap T''$ is failure-free. Then $(F' \cap P'') \cup (P' \cap F'') = \emptyset$;

but $F \subseteq F'$ and $P \subseteq P'$, so $(F \cap P'') \cup (P \cap F'') = \emptyset$, also. This is the failure set for $T \cap T''$. Hence, by Lemma 4.2, $T \preceq T'$.    □

We make heavy use of the following results relating quotient and concatenation.

**Lemma 4.7.**  *If* $X, Y \subseteq A^*$, $X/Y \subseteq X$ *iff* $(A^* - X)Y \subseteq (A^* - X)$.

**proof.**  $X/Y \subseteq X$ iff for every $x \in A^*$ and $y \in Y$, $xy \in X$ implies $x \in X$. $(A^* - X)Y \subseteq$ $(A^* - X)$ iff for every $x \in A^*$ and $y \in Y$, $x \notin X$ implies $xy \notin X$. But these statements are equivalent.    □

**Lemma 4.8.**  *For any* $X, Y \subseteq A^*$, $XY \subseteq X$ *iff* $XY^* \subseteq X$.

**proof.**  It is easy to show by induction that $XY^i \subseteq X$ for all $i \in \omega$, so $XY^* = X(\bigcup_{i \in \omega} Y^i) \subseteq X$.    □

**Lemma 4.9.**  $X/Y \subseteq X$ *iff* $X/Y^* \subseteq X$

**proof.**  Let $X' = (A^* - X)$. Then, $X/Y \subseteq X$ iff $X'Y \subseteq X'$ (Lemma 4.7) iff $X'Y^* \subseteq X'$ (by Lemma 4.8) iff $XY^* \subseteq X$.    □

**Lemma 4.10.**  $X \subseteq A^*$ *is prefix-closed if and only if* $X/A \subseteq X$.

**proof.**  It is obvious from the definition of prefix-closure that it is equivalent to $X/A^* \subseteq X$. This is equivalent to $X/A \subseteq X$ by Lemma 4.9.    □

The following theorem is the major result of this section:

**Theorem 4.2.**  *If* $T$ *is any prefix-closed trace structure, performing autofailure manifestation followed by failure exclusion yields a canonical prefix-closed trace structure that is conformation equivalent to* $T$.

The theorem is proved through several lemmas. First, we show that the results of the simplification are prefix-closed trace structures which are conformation equivalent to the originals:

**Lemma 4.11.**  *Autofailure manifestation preserves prefix-closed trace structures.*

**proof.**  We prove only the less obvious properties. $F'$ is regular because regular sets are closed under quotient and concatenation with regular sets. To see that $P$ is prefix-closed, let $x$ be any member of $P'$. If $x \in P$, every prefix of $x$ is in $P \subseteq P'$. If $x \in P' - P$, we know $x \in (F/O^*)A^*$. Then there is a prefix $y$ of $x$ in $F/O^*$ and some $z \in O^*$ such that $yz \in F$. So $yz \in P$, so every prefix of $yz$, including every prefix of $y$ is in $P \subseteq P'$. Also, since $y \in F$, $yA^* \subseteq F' \subseteq P'$. But this set includes every sequence $w$ in $y \leq w \leq x$, so every prefix of $x$ is in $P'$.    □

**Lemma 4.12.** *Autofailure manifestation preserves conformation equivalence.*

**proof.** Let $T$ be any prefix-closed trace structure, and let $T'$ be the result of auto-failure manifestation. By definition $F' = (F/O^*)A^*$ and $P' = S \cup F'$; all other aspects of $T$ and $T'$ are the same.

Clearly, $F \subseteq F'$ and $P \subseteq P'$, so, by Lemma 4.6, $T \preceq T'$. To see that $T' \preceq T$, let $T''$ be any trace structure such that $I'' = O$ and $O'' = I$. We show that if there is a failure in $T' \cap T''$, there is a failure in $T \cap T''$. Suppose that there is a failure $x$ in $T' \cap T''$. Either $x \in F' \cap P''$ or $x \in P' \cap F''$. If $x \in F' \cap P''$, we know $x \in (F/O^*)A^*$, so there is some prefix $z$ of $x$ in $F/O^*$ ($z$ is an autofailure), and there is a $y \in O^*$ such that $zy \in F$. Since $x$ is in $P''$, which is prefix-closed, $z \in P''$, also. $T''$ is receptive and $y \in O^* = I''^*$, so $zy \in P''I^* \subseteq P''$. Hence, $zy \in F \cap P''$; whenever $x$ is in $F' \cap P''$, $zy \in F \cap P''$, which is a subset of the failures of $T \cap T''$.

Otherwise $x \in P' \cap F''$ and $x \notin F' \cap P''$. Since $F'' \subseteq P''$, this implies $x \in (P' - F') \cap F''$. But, clearly, $P' - F' = P - F$, so $x \in (P - F) \cap F''$, which is a subset of the failures of $T \cap T''$. Hence, whenever there is a failure in $T' \cap T''$ there is a failure in $T \cap T''$; equivalently, whenever $T \cap T''$ is failure-free, $T' \cap T''$ is, too. So, by Lemma 4.2, $T' \preceq T$. Hence, $T' \sim_c T$.  $\square$

**Lemma 4.13.** *If $T$ is any prefix-closed trace structure such that $FA = F$, failure exclusion yields a prefix-closed trace structure.*

**proof.** Let $T'$ be the result of applying the transformation to $T$. The transformation changes nothing except the successes; we need only prove that $S' = S - F$ is regular and prefix-closed. It is regular since regular sets are closed under set complement and intersection. $S$ is prefix-closed, so by Lemma 4.10, $(A^* - S)A \subseteq A^* - S$. By the definition of $S'$ and de Morgan's law, $A^* - S' = (A^* - S') \cup F$. By hypothesis, $FA \subseteq F$. so

$$[(A^* - S') \cup F]A = (A^* - S)A \cup FA \subseteq [(A^* - S) \cup F)] = A^* - S.$$

Hence, by Lemma 4.10 again, $S'$ is prefix-closed.  $\square$

**Lemma 4.14.** *If $T$ is any prefix-closed trace structure and $T'$ is the result of applying autofailure manifestation, $T'$ satisfies $F'/O \subseteq F'$ and $F'A \subseteq F'$.*

**proof.** By definition, $F' = (F/O^*)A^*$. Let $F_1 = F/O^*$. By property 3.2, $(F/O^*)/O^* = F/(O^*O^*) = (F/O^*)$, so $F_1/O^* \subseteq F_1$. Hence $F_1/O \subseteq F_1$, by Lemma 4.9.

Let $F_2 = F_1A^*$. Obviously, $(F_1A^*)A \subseteq F_1A^*$, so $F_2A^* \subseteq F_2$. We still need to show that appending $A^*$ to $F_1$ does not destroy the first property. Let $x$ be any member of $F_2/O$. Then there is some $a \in O$ such that $xa \in F_2$. If $xa \in F_1$, we know $x \in F_1$, also, since $F_1/O \subseteq F_1$. Suppose, $xa \in F_1A^+$; since $a \in A$, also, this implies $x \in F_1A^* = F_2$.  $\square$

**Lemma 4.15.** *Failure exclusion preserves conformation equivalence.*

**proof.** Let $T$ be the any prefix-closed trace structure, and let $T'$ be the result of excluding failures in it. $P = P'$ and $F = F'$, so by two applications of Lemma 4.6, $T \sim_c T'$. □

Failure exclusion does not change $F$, so it preserves the properties $F/O \subseteq F$ and $FA \subseteq F$ established by autofailure manifestation. Obviously, after failure exclusion, $S \cap F = \emptyset$. This concludes the proof of the theorem.

The next lemma proves that the set of failures is a function of the successes in a canonical prefix-closed trace structure. Given a trace structure $T$,

**Lemma 4.16.** *For any canonical prefix-closed trace structure $T$, $F = [(SI \cup \{\epsilon\}) - S]A^*$.*

**proof.** Define $F_0$ to be the minimal sequences in $F$; in other words,

$$F_0 = \{x \mid x \text{ has no proper prefixes in } F\}.$$

If $S = \emptyset$, $\epsilon \in F$ because $\epsilon \in P = S \cup F$. Since $\epsilon$ is a prefix of everything, it must be the only minimal element of $F$, so $F_0 = \{\epsilon\} = (SI \cup \{\epsilon\} - S)$ in this case.

Otherwise, $S \neq \emptyset$. Let $x$ be any member of $F_0$. Since $S$ is non-empty, it has prefixes in common with $x$ ($\epsilon$, if nothing else) so $x$ has a maximum prefix in $S$, which we call $y$. Hence, $x = yz$ where $z \in A^*$. We prove that $|z| = 1$. If $|z| = 0$, ($z = \epsilon$), we would have $x = y \in S$, which is a contradiction because $x \in F$ and $S \cap F = \emptyset$. If $|z| > 1$, then $z = az'$ where $z' \neq \epsilon$; then $ya \in F$, since $y$ is the *maximum* prefix of $x$ in $S$. But $ya$ would a proper prefix of $x$ in $F$, which contradicts $x \in F_0$. Hence, the only remaining possibility is $|z| = 1$, or equivalently, $z \in A$. Hence, $x \in SA - S$.

Now $x \notin SO$, because otherwise it would have a proper prefix that was an autofailure and would not be in $F_0$. Hence, $x \in SI - S$, so $F_0 \subseteq SI - S$. Furthermore, $SI \subseteq P$ by receptiveness, so $SI - S \subseteq P - S = F$. Since every proper prefix of every member of $SI - S$ is a success, $SI - S \subseteq F_0$. Hence, $F_0 = SI - S$, and since $\epsilon \in S$, $(SI \cup \{\epsilon\}) - S = SI - S$.

Hence, in either case, $F_0 = (SI \cup \{\epsilon\} - S)$. By the definition of $F_0$, $F \subseteq F_0A^*$. But $T$ is canonical, so $FA^* \subseteq F$, so $F_0A^* \subseteq FA^* \subseteq F$. Hence $F = F_0A^* = [(SI \cup \{\epsilon\} - S)]A^*$. □

Although it is obvious that every canonical trace structure can be converted to a simple one (by discarding the $F$ set), it is not as clear that every simple prefix-closed trace structure can be converted to a canonical one. It could be that there are prefix-closed $S$ sets which appear in no canonical trace structures. The next lemma shows that this is not a problem.

**Lemma 4.17.** *If I and O are disjoint finite sets of wires, A = I ∪ O, and S is any regular prefix-closed subset of A\*, then (I, O, S, [(SI ∪ {ε}) − S]A\*) is a prefix-closed trace structure.*

**proof.** The inputs, outputs, and successes satisfy all the necessary conditions by hypothesis. $P$ is obviously regular. To see that it is prefix-closed, let $x$ be any member of $P$. If $x \in S$, all its prefixes are in $S$, also, so they are all in $P$. Otherwise, $x \in [(SI \cup \{\epsilon\}) - S]A^*$, so there is a $y \in [(SI \cup \{\epsilon\}) - S]$ and $z \in A^*$ such that $x = yz$. Every prefix of $y$ is in $P$ because it is either $y$ itself, or in $S$, or it is $\epsilon$. Clearly, if $w$ is a prefix of $z$, $yw \in [(SI \cup \{\epsilon\}) - S]A^*$, too. This accounts for all prefixes of $x$. For receptiveness, let $x$ be any member of $P$ and $a$ be any member of $I$. If $x \in S$, either $xa \in S$ or $xa \in SI - S \subseteq F$. If $x \in [(SI \cup \{\epsilon\}) - S]A^*$, so is $xa$, obviously. So in any case $xa \in P$, hence $PI \subseteq P$.  □

## 4.4.  Environments and Mirrors

The primary result of this section is that for almost every $T$ there is a unique maximum environment that can be composed with $T$ failure-free. This result gives an exact characterization of the environments allowed by a trace structure $T$: they are the environments that conform to the maximum one. This result has several important implications: it gives a simple a decision procedure for conformation. It is an important tool for proving further results; for example, this result makes it easy to prove that two canonical prefix-closed trace structures are equal if and only if they are conformation-equivalent.

To find the maximum environment of a trace structure $T$, $T$ must first be reduced to canonical form using the simplifications of the previous section. The maximum environment is called the *mirror* of $T$, (written $T^M$). If $T$ is canonical, $T^M$ has $I^M = O$, $O^M = I$, $S^M = S$, and $F^M = A^* - P$. It follows that $P^M = A^* - F$. The mirror is also a canonical prefix-closed trace structure. We said above "almost every $T$." The exceptions are the structures that have $S = \emptyset$. Since $P = A^*$, the mirror would have $P^M = S \cup (A^* - P) = \emptyset$, which is not a trace structure ($P$ is empty). There is exactly one such trace structure for each pair of input and output sets.

It follows immediately from the result of the previous paragraph that is surprisingly easy to define the mirror for *simple* prefix-closed trace structures: just exchange the inputs and outputs. Once again, the mirror is not defined in case $S = \emptyset$ — the inputs and outputs can be exchanged, but the result is not helpful.

Mirrors can be used in a simple test for conformation. Instead of checking the compositions of $T$ and $T'$ with all other trace structures (an imposing task), it is only necessary to compose with one: the mirror of $T'$. If $T \parallel T'^M$ is failure-free, $T \preceq T'$. Intuitively, $T'^M$ is the worst possible environment that the specification can be composed with; it accepts everything that $T'$ can produce as output and nothing more, and produces everything that $T'$ can accept as input and nothing less. In a sense, $T'^M$ is a conceptual implementation-tester. Deciding whether the composition of two trace structures is failure-free requires a few straightforward operations on finite automata, so this gives a convenient *decision procedure* for $T \preceq T'$. This test fails in case $T'$ has an empty success set, because $T'^M$ is not defined. However, this special case can be checked separately.

The results of this section allow two claims of the previous section to be proved: first, two canonical prefix-closed trace structures are equal if they are conformation-equivalent; second, canonical prefix-closed trace structures are a circuit algebra.

The idea of composing the mirror of a specification with the implementation has been published by Ebergen ([30,31]). However, the justification in terms of failures, substitutions, and conformation are new.

We now prove the claims of this section.

**Lemma 4.18.** *If $T$ is a canonical prefix-closed trace structure and $S \neq \emptyset$ then $T^M$ is a canonical prefix-closed trace structure.*

**proof.** The only non-obvious conditions for $T^M$ to be a trace structure are those on $P^M$: non-emptiness, prefix-closure, and receptiveness. Non-emptiness is immediate, because $S$ is non-empty by hypothesis. Since $T$ is canonical, $FA \subseteq F$, so by Lemma 4.7,

$$P^M/A \subseteq (A^* - F)/A \subseteq (A^* - F) = P^M,$$

so $P^M$ is prefix-closed by Lemma 4.10.

$F/O \subseteq F$ because $T$ is canonical, and by Lemma 4.7, $(A^* - F)O \subseteq (A^* - F)$. Since $P^M = A^* - F$, and $I^M = O$, this proves $P^M I^M \subseteq P^M$ — $T^M$ is receptive.

To prove $T^M$ is canonical, simply reverse these arguments. Since $T$ is receptive, $PI \subseteq P$, which implies $F^M/O^M \subseteq F^M$. $P$ is prefix-closed, so $P/A \subseteq P$, so $F^M A \subseteq F^M$.
□

**Theorem 4.3.** *For any simple prefix-closed trace structures $T$ and $T'$ $T \preceq T'$ iff $I = I'$, $O = O'$, and either $S' = \emptyset$ or $T \cap T'^M$ is failure-free.*

This theorem is proved in the next few lemmas.

**Lemma 4.19.** *If $T$ and $T'$ are simple trace structures such that $I = I'$, $O = O'$, and $S' = \emptyset$, then $T \preceq T'$.*

**proof.** The composition of $T'$ with any trace structure is never failure-free, so the antecedent of Lemma 4.2 is trivially satisfied.  □

**Lemma 4.20.** *For any simple prefix-closed trace structures $T$ and $T'$ such that $I = I'$, $O = O'$, and $S' \neq \emptyset$, if $T \cap T'^M$ is failure-free, $T \preceq T'$.*

**proof.** Suppose $T \cap T'^M$ is failure-free. Then $P \cap F'^M = \emptyset$ and $F \cap P'^M = \emptyset$. Substituting the definition of $P'^M$, we have $P \cap (A^* - P') = \emptyset$, so $P \subseteq P'$; similarly, $F \subseteq F'$. So by Lemma 4.6, $T \preceq T'$.  □

**Lemma 4.21.** *If $S \neq \emptyset$, $T \cap T^M$ is failure-free.*

**proof.** Let $T' = T \cap T^M$, so $F' = (F \cap P^M) \cup (P \cap F^M)$. But $F \cap P^M = F \cap (A^* - F) = \emptyset$ and $P \cap F^M = P \cap (A^* - P) = \emptyset$.  □

**Lemma 4.22.** *If $T \preceq T'$ and $S' \neq \emptyset$, $T \cap T'^M$ is failure-free.*

**proof.** By Lemma 4.21, $T' \cap T'^M$ is failure-free, so by Lemma 4.2, $T \cap T'^M$ is failure-free, also.  □

Lemmas 4.19, 4.20, and 4.22 prove Theorem 4.3.

Lemma 4.6 says that it is *sufficient* for $T \preceq T'$ that $P \subseteq P'$ and $F \subseteq F'$. For simple prefix-closed trace structures, this condition is also *necessary*.

**Lemma 4.23.** *If $T$ and $T'$ are any canonical prefix-closed trace structures and $T \preceq T'$, then $I = I'$, $O = O'$, $P \subseteq P'$, and $F \subseteq F'$.*

**proof.** $A' = A$ by hypothesis. If $S' = \emptyset$, every trace structure $T$ with the same inputs and outputs conforms to $T'$: $\epsilon \in P$, so $\epsilon \in F$. Since $T'$ is canonical, $F'A^* \subseteq F'$, so $\epsilon A^* = A^* \subseteq F' \subseteq P'$. By the definition of trace structures, $F' \subseteq P' \subseteq A^*$, so $F' = P' = A^*$. Therefore, $P \subseteq A^* = P'$ and $F \subseteq A^* = F'$.

If $S' \neq \emptyset$, $T'^M$ is defined. In this case, by Lemma 4.22, $T \cap T'^M$ is failure-free. The set of failures is $(F \cap P'^M) \cup (P \cap F'^M)$, so $F \cap (A^* - F') = \emptyset$ and $P \cap (A^* - P') = \emptyset$. But then $F \subseteq F'$ and $P \subseteq P'$.  □

We now prove the two results from the previous section that were deferred until now. First, canonical prefix-closed trace structures are a canonical form with respect to conformation equivalence.

**Lemma 4.24.** *If $T$ and $T'$ are any canonical prefix-closed trace structures, $T \sim_C T'$ iff $T = T'$.*

**proof.** Trivially, $T = T'$ implies $T \sim_C T'$. Also, if $I \neq I'$ or $O \neq O'$, neither $T \sim_C T'$ nor $T = T'$.

Now suppose $T \sim_C T'$. By Lemma 4.23, $P \subseteq P'$ and $P' \subseteq P$ so $P = P'$. Similarly, $F = F'$. Since $P = S \cup F$ and $S \cap F = \emptyset$, $S = P - S = P' - S' = S'$.    □

The second deferred proof was that canonical prefix-closed trace structures are a circuit algebra. Recall that the operations on general trace structures were modified to simplify their results to canonical form. Note that, by Lemma 4.24, canonical trace structures are isomorphic to the equivalence classes of trace structures under conformation equivalence. We prove that conformation is a congruence for circuit algebra; hence, the quotient of the circuit algebra of trace structures with respect to conformation equivalence is also a circuit algebra, and so are the canonical trace structures.

**Lemma 4.25.** *Conformation equivalence is a congruence for circuit algebra.*

**proof.** We need to show that if $T_i \sim_C T_i'$ for $i \in \{1, 2\}$, $(T_1 \parallel T_2) \sim_C (T_1' \parallel T_2')$, **hide**$(D)(T_1) \sim_C$ **hide**$(D)(T_1')$, and **rename**$(r)(T_1) \sim_C$ **rename**$(r)(T_1')$. These results follow immediately by applying Theorem 4.1 in two directions.    □

## 4.5.  Examples

This section has several examples to illustrate the $\preceq$ relation.

### 4.5.1. Merge Element

A *merge element* is a standard building block used in self-timed systems. It has two inputs and an output (Figure 4.3a). If there is a transition on either input, a transition is signaled on the output. No input transitions (on either input) are allowed between the input and output transition. An automaton for a merge element is shown in Figure 4.3b. An XOR gate is a merge element.

Consider a merge that requires its inputs to be of the form "*acbcacbc*....". We call this an *alternating merge* (it is not particularly useful except as an example). An automaton for a alternating merge element is shown in Figure 4.3c. Let the trace structure for the first merge be $T_{GM}$ (for "general merge") and let the second be $T_{AM}$.

$F_{GM}$, the set of failures of the general merge, consists of traces where two or more inputs occur without an intervening output. $F_{AM}$ includes $F_{GM}$ and has additional traces of the form "*acbc...acacA\**" and "*acbc...bcbcA\**", so $F_{GM} \subset F_{AM}$. Similarly, $P_{GM} \subset P_{AM}$: if a trace $x$ in $P_{GM}$ is a failure, $x$ is in $P_{AM}$, also, because $F_{GM} \subset F_{AM}$. If $x$

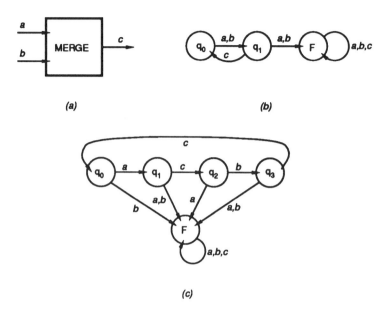

**Figure 4.3.** Merge Element and Alternating Merge

is a success, it is either in $S_{AM}$, or $x$ is of the form "$acbc\ldots acac\ldots$" or of the form "$acbc\ldots bcbc\ldots$", in which case it is in $F_{AM}$. So by Lemma 4.6, $T_{GM} \preceq T_{AM}$.

Returning to intuition, a general merge can be safely substituted for an alternating merge in any context because a general merge can handle any input that an alternating merge can handle (and some more), and produces exactly the same outputs when given those inputs.

However, $T_{AM}$ does not conform to $T_{GM}$: substituting an alternating merge for a general merge can introduce failures into previously failure-free contexts. This occurs when the environment sends inputs in non-alternating order — the general merge can handle them, but the alternating merge cannot.

### 4.5.2. Arbitrary Selector

A *selector* is the mirror of a merge element. Whenever it receives a transition on its only input, a transition is signaled on one of two outputs. A general selector can signal either of the two outputs arbitrarily, while an alternating selector must signal them in alternating order. The success sets for these two circuits are exactly the same as for the corresponding merge elements, but the input and output alphabets are interchanged. Let $T_{GS}$ be the trace structure for a general selector and $T_{AS}$ be the alternating selector.

In this case, $T_{AS} \preceq T_{GS}$. This is *exactly the opposite* of the situation with the merge elements. $F_{GS}$ consists of the traces in which two or more inputs occur without an output, as does $F_{AS}$. However, $F_{GS}$ has traces of the form "$\dots acac \dots acbcc \dots$" that $F_{AS}$ does not (because the successful prefixes of $F_{AS}$ have alternating outputs), so $F_{AS} \subset F_{GS}$; $P_{AS} \subset P_{GS}$ for the same reason.

Intuitively, $T_{AS}$ can be safely substituted for $T_{GS}$ because it is less likely to break the environment — if the environment can handle the outputs of $T_{GS}$, it can certainly handle the outputs of $T_{AS}$. Also, $T_{AS}$ can handle the same inputs as $T_{GS}$, given the outputs it produces.

However, $T_{GS}$ cannot be safely substituted for $T_{AS}$ because it may produce outputs that cause failures in the environment, when $T_{AS}$ would not have done so. Hence, $T_{GS} \npreceq T_{AS}$.

### 4.5.3. Set-Reset Latch

A latch was informally specified in Section 3.4; here, we give a trace structure as an equivalent formal specification. An automaton for the success set of the specification is shown in Figure 4.4a. This is a simple trace structure, so failures can be computed from the success set (the failures are not shown in the diagram). The specification assumes that $a$ and $b$ are never simultaneously high. It also allows the outputs to change between the set and reset states in either order ($cd$ or $dc$). This is one of several possible specifications; another appears in Chapter 5.

The automaton for the success set of the simple trace structure for the cross-wired NOR of the previous chapter is shown in Figure 4.4b. The state 0000 was removed from the success set because it was an autofailure. This example has a mixture of the properties of the merge and selector examples. For example, in state 0101, the implementation accepts a superset of the inputs that the specification requires because it can accept an $a$. The implementation also produces a subset of the outputs allowed by the specification because it is allowed to generate a $c$ in state 1001 but it does not. The implementation conforms to the specification but the reverse is not true.

### 4.5.4. Fork/Join

A *fork/join* module (Figure 4.5a is used to spawn two parallel processes. When the fork/join receives a transition on $r$, it sends transitions on $lr$ and $rr$ to start two sub-computations. When these computations complete, they send signals $la$ and $ra$, respectively, back to the fork/join. When both of these transitions have been received by the fork/join, it signals the completion of the entire computation by a transition on $a$.

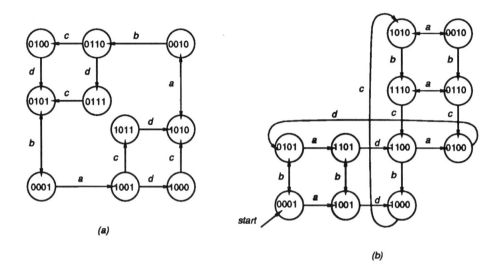

**Figure 4.4.** Specification and Implementation of Set-Reset Latch


Figure 4.5b is a state diagram for a fork/join. The two subprocesses are signaled in any order, and run concurrently, so the 'finished' signal from one process can occur in any order relative to the 'start' signal for the other.

Figure 4.5c shows a state diagram for a 'sequence' element, which starts one process and waits for it to complete before starting the other. The sequence module conforms to the specification of the fork/join. Intuitively, it produces a subset of the allowed outputs after the traces $r$ and $r; lr; rr$, and accepts all the specified inputs after the prefix $r; ra$, the prefix $r; lr; la; rr$, and the prefix $r; lr; la; rr; ra$.

This result illustrates a limitation of the expressive power of prefix-closed trace structures: there is no way to specify a *concurrent* implementation of the fork/join. There are good reasons to require concurrent implementations of circuits. The most obvious is speed; the combined completion time of two concurrent processes is the maximum of the completion times of the two processes, while the completion time of two sequential processes is the sum of their individual times. But real-time performance requirements cannot be expressed using trace structures.

Another reason for concurrent implementations is liveness. Intuitively, an environment could distinguish the fork/join and sequence by withholding the $d$ signal after the trace

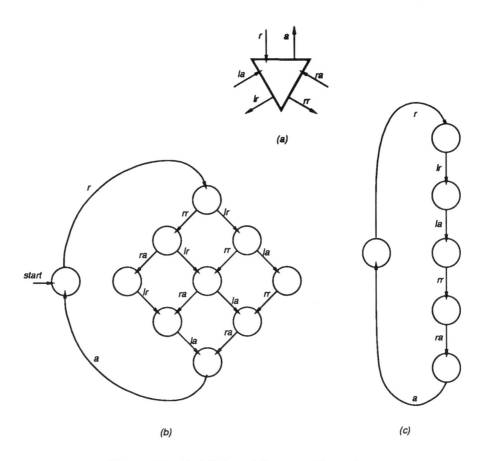

**Figure 4.5.** Fork/Join and Sequence Elements

*ab*. If the fork/join were live, it would eventually produce a *c*, anyway. However, the sequence module would wait forever in this scenario. But prefix-closed trace structures cannot specify liveness properties, either. We will see in Chapter 7 that a concurrent implementation *can* be specified using complete trace structures because they can express liveness properties.

### 4.5.5. Universal Do-Nothing Module

Molnar has proposed a 'Universal Do-Nothing Module', a circuit that accepts all inputs and produces no outputs. This device can be manufactured out of wood. The trace structure for a universal do-nothing module with inputs $I$ and outputs $O$ is $(I, O, I^*)$. This module satisfies any specification having the same alphabet, which illustrates graphically that prefix-closed trace structures specifications cannot require a circuit to *do* anything.

## 4.6.   Delay Insensitivity

Delay insensitivity has been a central concern of previous work in trace theory. A trace structure is delay-insensitive if adding delays to its inputs and outputs does not change it. An implementation of a delay-insensitive specification is guaranteed to work properly no matter what wire delays are introduced in realizing the circuit.

Unlike Snepscheut [72,73] and Black [9], we do not require that circuits be delay-insensitive in order for our composition operator to be valid. However, delay insensitivity is a desirable property, which we would like to be able to check automatically. This section characterizes delay insensitivity using the results of this chapter. This approach allows a very simple definition of delay insensitivity, and provides a simple decision procedure to test it, too.

We define an operator on trace structures called **DI**, which finds the least delay-insensitive specification that the original specification conforms to. A delay is equivalent to a non-inverting buffer (identity gate). **DI** attaches non-inverting buffers on all the inputs and outputs of a trace structure, then hides the original wires and renames the new inputs and outputs to the original names. The overall effect of **DI** is to remove some ordering relations between signals. For example, if $T$ outputs $ab$ but not $ba$, $T' = \mathbf{DI}(T)$ outputs both $ab$ and $ba$, because signals may emerge from delays in a different order than they entered. Thus, **DI** can enlarge the $P$ and $F$ sets. In fact, $P \subseteq P'$ and $F \subseteq F'$, since signals may also emerge from delays in exactly the same order they entered. Hence, for all trace structures, $T \preceq \mathbf{DI}(T)$.

**DI** can be defined formally as follows: given a trace structure $T$, choose alphabets $I'$ and $O'$ disjoint from each other and from $I$ and $O$, and a renaming function $r: (I' \cup O') \to A$. For each $a \in I$, let $T_a$ be the trace structure resulting from renaming the input of a non-inverting buffer to be $r^{-1}(a)$ and the output to be $a$; similarly, for each $b \in O$, let $T_b$ be a buffer renamed so the input is $b$ and the output is $r^{-1}(b)$. Then define

$$\mathbf{DI}(T) = \mathbf{rename}(r)[\mathbf{hide}(A)(T_a \parallel \ldots \parallel T_b \parallel \ldots T)].$$

Delay insensitivity can be defined very easily using the **DI** operator: a trace structure $T$ is delay-insensitive if and only if $\text{DI}(T) \preceq T$. Since the effect of **DI** can be computed using the operations of circuit algebra and conformation is decidable, delay insensitivity is decidable.

Composing *two* delays and hiding the connection between them yields a trace structure that is conformation-equivalent to the original. Hence, **DI** is *idempotent*: for every $T$, $\text{DI}[\text{DI}(T)] = \text{DI}(T)$. Adding *more* delays to the inputs and outputs of a delay-insensitive circuit has no effect on its operation.

## 4.7. Lattice Properties

One of the advantages of logical formulas is that specifications can be combined using conjunction, disjunction, and negation. For example, to verify that a system satisfies the logical formulas $\phi \wedge \psi$, it may be possible to show that it satisfies $\phi$ and $\psi$ separately, then infer from this that it satisfies both simultaneously.

Trace structures can be combined using operators similar to $\vee$ and $\wedge$. There is no operation equivalent to $\neg$, because safety properties are not closed under negation. If $T$ and $T'$ are two trace structures with matching alphabets, the *meet* of $T$ and $T'$ (written $T \sqcap T'$) is the "most general" trace structure that conforms to both $T$ and $T'$. Similarly, the *join* of $T$ and $T'$ (written $T \sqcup T'$) is the "least general" trace structure such that both $T$ and $T'$ conform to it. More formally, the set of canonical prefix-closed trace structures having input and output alphabets $I$ and $O$ form a bounded distributive lattice under the $\preceq$ ordering.

The maximum element of the lattice is $(I, O, \emptyset)$. This is the universal specification — everything conforms to it, and it conforms to nothing but itself. The minimum element is $(I, O, I^*)$. This is the universal implementation (also known as the "universal do-nothing module"), which conforms to everything.

The meet and join operations may be helpful in breaking up a verification task into smaller parts. For example, if $T \preceq T'$ and $T \preceq T''$, then $T \preceq (T' \sqcap T'')$ — if a circuit can be shown to satisfy two partial specifications, it satisfies both of them simultaneously.

There is a simple way to compute these operations on canonical prefix-closed trace structures, based on Lemmas 4.6 and 4.23. If $T$ and $T'$ have matching alphabets, $T'' = T \sqcap T'$ is the canonical trace structure having $P'' = P \cap P'$, $F'' = F \cap F'$, and $S'' = P'' - F''$. Similarly $T'' = T \sqcup T'$ is the canonical trace structure having $P'' = P \cup P'$, $F'' = F \cup F'$, and $S'' = P'' - F''$.

The mirror operation behaves somewhat like a complementation operation, although it is *not* a complementation in the lattice because it changes the inputs and outputs of its operand. There are some interesting interactions between the lattice operations and mirroring, which are reminiscent of de Morgan's laws:

$$(T \sqcap T')^M = T^M \sqcup T'^M, \text{ and}$$
$$(T \sqcup T')^M = T^M \sqcap T'^M.$$

There is a possibility that $(T \sqcap T')^M$ is defined when $T$ or $T'$ is the universal specification (for which the mirror is not defined). However, $(T \sqcup T')$ can be the universal specification only if $T$ or $T'$ is (because $\epsilon \in F \cup F'$ only if $\epsilon \in F$ or $\epsilon \in F'$), so the two sides of the second identity are either both defined or both undefined.

Meet and join can be used to add explicit environmental assumptions to a specification. Suppose we are interested in composing a trace structure $T$ only with environments that conform to $T'$. By itself, $T$, has an implicit restriction that its environment must conform to $T^M$. The additional requirement says that any environment must also conform to $T'$; so it must conform to $T^M \sqcap T'$. The desired trace structure is the maximum that can be composed with $T^M \sqcap T'$ failure-free; in other words, the maximum environment of the environment, $(T^M \sqcap T')^M = T \sqcup T'^M$. Adding environmental assumptions can simplify a trace structure: aspects of its behavior that were relevant in the larger environment can be suppressed if they are irrelevant in the restricted environment.

The rest of this section is devoted to proofs of these claims. Formally, we define $T'' = T \sqcap T'$ when $I = I'$ and $O = O'$ so that $T'' \preceq T$ and $T'' \preceq T'$ and for every $T'''$ satisfying the same conditions, $T''' \preceq T''$. Similarly, if $T'' = T \sqcup T'$, then $T \preceq T''$ and $T' \preceq T''$ and for every $T'''$ satisfying the same conditions, $T'' \preceq T'''$.

**Theorem 4.4.** *Given disjoint finite alphabets $I$ and $O$, the set of all canonical prefix-closed trace structures having $I$ and $O$ as their input and output alphabets is a bounded distributive lattice under the conformation ordering.*

In the remainder of this section, let $T$ and $T'$ be any canonical prefix-closed trace structures such that $I = I'$ and $O = O'$. There are dual versions of each result, with meet and join interchanged. Since the proofs are generally quite similar, we prove only one of the cases. We prove that $T \sqcap T'$ and $T \sqcup T'$ are well-defined by showing how to compute them.

**Lemma 4.26.** $T'' = (I, O, (P \cap P') - (F \cap F'), F \cap F')$ *is a canonical prefix-closed trace structure.*

**proof.** Note that $P'' = P \cap P'$. It is easy to confirm that $T''$ is a prefix-closed trace structure. That it is canonical remains to be proved.

Since $T$ and $T'$ are canonical, $F/O \subseteq F$ and $F'/O \subseteq F'$, so by property 3.1,

$$F''/O = (F \cap F')/O \subseteq F/O \cap F'/O \subseteq F \cap F' = F''.$$

Also, $FA \subseteq F$ and $F'A \subseteq F'$, so $F''A = (F \cap F')A \subseteq FA \cap F'A \subseteq F \cap F' = F''$. Obviously, $S'' \cap F'' = \emptyset$, so $T''$ is canonical. □

**Lemma 4.27.** $T \sqcap T' = (I, O, (P \cap P') - (F \cap F'), F \cap F')$.

**proof.** Let $T'' = (I, O, (P \cap P') - (F \cap F'), F \cap F')$. So by Lemma 4.6, $T'' \preceq T$ and $T'' \preceq T'$. Now suppose that $T''' \preceq T$ and $T''' \preceq T'$. Then by 4.23, $P''' \subseteq P \cap P'$ and $F''' \subseteq F \cap F'$. By Lemma 4.6, again, $T''' \preceq T''$. □

The next lemma shows that the lattice is distributive.

**Lemma 4.28.** *Meet and join distribute over each other.*

**proof.** Follows from the distributive laws of union and intersection. □

Theorem 4.4 follows from Lemmas 4.26 through 4.28; the bounds of the lattice are the universal specification and the universal implementation.

Finally, we prove that mirroring interchanges meet and join.

**Lemma 4.29.** *For any canonical prefix-closed trace structures $T$ and $T'$, $(T \sqcap T')^M = T^M \sqcup T'^M$*

**proof.** Let $T_1 = (T \sqcap T')^M$ and $T_2 = T^M \sqcup T'^M$. $P_1 = A^* - (F \cap F') = (A^* - F) \cup (A^* - F') = P_2$ and $F_1 = A^* - (P \cap P') = (A^* - P) \cup (A^* - P') = F_2$ by de Morgan's laws. So $T_1 \sim_c T_2$, by Lemma 4.6. Hence, by Lemma 4.24, $T_1 = T_2$. □

# Chapter 5

# An Automatic Verifier

## 5.1. Introduction

This research was motivated by a practical problem: the difficulty of designing correct speed-independent circuits. The purpose of this chapter is to demonstrate that automatic verification is not only possible in theory, but practical as well. The ideas of the previous chapters have been implemented in a program that can quickly verify or find bugs in speed-independent circuits of nontrivial size. It has discovered errors in published designs.

The program is a "quick-and-dirty" prototype, which was written to demonstrate the feasibility of the method with as little programming effort as possible; the entire program consists of about 1,500 lines of LISP code. The program works well despite its simplicity, which attests to the value of trace theory as a basis for practical verification.

The program was written in Spice LISP (called LISP henceforth), a dialect of Common LISP [74]. LISP was chosen because it is well-suited for writing prototypes rapidly. It is particularly easy to embed *ad hoc* description languages in LISP. This feature was used in the program to facilitate the definition of state machines and circuit algebra expressions. One particular benefit is that the full power of LISP is available within the description language. For example, trace structures and circuit algebra expressions can be stored in variables (using the SETQ function) and abbreviations can be defined using functions and macros. The program was implemented and the examples run on a PERQ workstation.

Incidentally, LISP code appears in several places below. A knowledge of LISP is *not* necessary to understand this material — everything is explained in English. The LISP descriptions are included for concreteness; they are exactly what were used to verify the examples.

The next two of the following sections describe the verifier. The second section covers representational issues: the data structures representing trace structures and circuit algebra expressions, and the facilities to help the user construct them. The third section describes the algorithm for checking conformation. Through the use of some simple tricks, the verifier can avoid building a complete trace structure describing the implementation — a savings that is crucial in many cases.

The fourth and fifth sections describe examples, and show how the verifier can be applied to them. Both examples are schemes for building arbiters of any size by connecting a collection of standard cells in a regular way. In each case, the circuits are verified at two levels of abstraction: a high level, at which it is verified that larger arbiters can be built from the cells; and a low level, at which it is verified that a circuit composed of more primitive components implements a single cell. The trace structure for an individual cell is used as a description of a primitive at the high level and as a specification of the desired behavior at the lower level.

The final section describes some additional capabilities of the program. Besides checking conformation, the program can interactively perform the operations of circuit algebra and reduce the results to canonical form. For example, the program can examine the results of applying the **DI** operator (of Section 4.6) to an arbitrary trace structure.

## 5.2.   Descriptions and Specifications

To verify a circuit, a user supplies a description of the implementation and a specification and asks the program to check whether the former conforms to the latter. The description of the implementation has two parts: a trace structures for the primitive components, and a circuit algebra expression describing how they are connected. The specification is also a trace structure.

It greatly simplifies the implementation of the program, especially the decision procedure for conformation, to keep trace structures in canonical form. Of course, some details of the behavior of circuits are lost in the reduction to canonical form; for example, the difference between a circuit that *might fail* on a particular input, and a circuit that *will fail* on that input. In general, the distinctions in behavior represented by general trace structures are not very interesting to circuit designers, who are interested in designing correct circuits. In this case, the canonical form keeps exactly the right amount of detail; any more would be confusing as well as computationally inefficient.

### 5.2.1. Representation of Prefix-Closed Trace Structures

An important result of the previous chapter is that canonical prefix-closed trace structures are completely determined by a single trace set: the set of successes. The reduced representation, which has only a set of inputs, a set of outputs, and a set of successes, is called a simple prefix-closed trace structure. This is essentially the representation used in the program. The structure representing them is called a SPCTS (an acronym that is admittedly unpleasant to pronounce). The fields are:

names, a vector of wire names, representing the set $A$. No wire names are repeated, so this establishes a one-to-one correspondence between the wire names and the numbers $0 \ldots k - 1$, where $|A| = k$.

$I$, a bit vector representing the set of input wires. The $i$th bit is 1 iff the wire in the $i$th position in names is an input wire.

$O$, a bit vector of output wires, with the same interpretation as $I$.

$S$, an array representing a deterministic state graph.

start, a number representing the start state of $S$.

The state graph is an $n \times k$ array, where $n$ is the number of states and $k$ is the size of the alphabet. The states are numbered 0 to $n - 1$. $S$ represents the transition function of a deterministic finite-state automaton: $S[i, j]$ is the successor when a transition is taken from state $i$ on symbol $j$ (the $j$th wire in names).

To define a full-fledged finite automaton, the initial state and the accepting states need to be designated. The start state is given by the start field of the SPCTS. Every state in the state graph is an accepting state, since the set of successes is prefix-closed. As was noted in the previous section, the set of failures is a function of the set of successes, so it does not need to be represented explicitly.

It is clear from the complexity of the diagrams for the relatively simple examples in Sections 3.4 and 4.5 that there needs to be some more abbreviated way to define sets of traces. This raises the issue of description languages, which has been declared to be beyond the scope of the thesis. In the following, some simple functions and macros are used to facilitate the definition of trace structures. The result is probably not a generally useful or desirable notation, but it suffices for our purposes.

### 5.2.2. Defining Boolean Gates

Boolean gates appear frequently in circuit designs. The Boolean function that defines a gate is a much more readable and succinct description than the automaton accepting the traces of the gate. Because of this, there is a macro called GATE that constructs the trace structure for a gate. GATE takes as inputs a list of input wire names, an output wire name, a LISP expression for the Boolean function (LISP has built-in Boolean operations on bitvectors), a list of initial logical values for the inputs and output (to describe the initial state), and returns a trace structure (a SPCTS) for the gate. The construction of the trace structure follows the definition of trace structures for gates in Section 3.4 almost exactly.

For example, let us describe the gate that has inputs $a$, $b$, and $c$, output $d$, starts in the state in which $a = 0$, $b = 0$, $c = 1$, and $d = 1$, and computes the Boolean function $d = (a \land \neg b) \lor \neg c$:

```
(GATE (a b c) d
      (logior (logand a (lognot b)) (lognot c))
      ((a 0) (b 0) (c 1) (d 1)))
```

Since some particular gates are heavily used, the program also provides specific functions to define them. For example, there is a macro to define a two-input AND gate, called AND2. The AND2 macro takes two input names, an output name, and a description of the initial wire values, and returns the appropriate SPCTS. To describe the AND gate with inputs $a$ and $b$ and outputs $c$ with initial values $a = 0$, $b = 1$, and $c = 0$, we write:

```
(AND2 a b c ((a 0) (b 1) (c 0))).
```

Of course, users can write their own lisp functions for abbreviations, as well.

### 5.2.3. Defining State Machines

There is a another macro, called SM, to help define more general circuit elements and higher-level specifications. The SM macro takes arguments to define the input and output alphabets of the trace structure, a state graph, and the initial state of the state graph, and returns a SPCTS. The most interesting part of the macro is the declarative description of a state machine. The states of the state graph are decomposed into a collection of named *state variables*. Each state variable has a declared range of values, from 0 to some (usually very small) positive integer. The transition function is defined symbolically by a set of clauses. Each clause has a *predicate* that is satisfied by a set of pairs of states and wires. Following the predicate, the clause has any number of *actions* that define the target state when the predicate is satisfied, by assigning new values to the state variables. The new values are in general a function of the old values of the variables. If no predicate is satisfied by a particular state/wire pair, there is no corresponding transition. A transition array is constructed from the clause by the brute-force method of iterating over all possible combinations of state variable values and wire names. When a predicate is satisfied by a state/wire pair, $(i, j)$, the next state is computed and placed in the $(i, j)$th entry of the array.

The examples of this section (and speed-independent designs, in general) make use of *hand-shaking protocols*. The one used here is *four-phase* signaling. This protocol uses two wires for each interface, a *request* wire and an *acknowledge*. One cycle of four-phase signaling consists of a request (transition), an acknowledge, another request, and another acknowledge. The initial request and acknowledge transition are usually interpreted as requesting a service and acknowledging that it will be or has been performed. Sometimes meanings are associated with the second request and acknowledge, also; other times, they exist solely to return the wires to logical 0 values for convenience in the implementation. A sequence diagram for this protocol is shown in Figure 5.1.

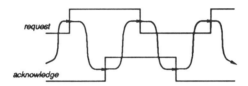

**Figure 5.1.**   Four-phase Signaling

As an example, consider the mutual exclusion element of Section 3.4 (Figure 3.4). It can be considered to use four-phase signaling. Their are two *users*. The request and acknowledge wires for the first user are *ur1* and *ua1*, and for the second are *ur2* and *ua2*. The user asks for the resource by raising the request wire. The arbiter cell grants the resource to the user by raising the acknowledge. The user then does whatever it wants with the resource, fully confident that none of the other users will be granted access. When the user is done, it releases the resource by lowering the request line. This is acknowledged by the arbiter by lowering the acknowledge line, so that the interface is ready for another user request.

Each interface of the ME element has four states, which are numbered 0 through 3. State number 0 is the *quiescent* state; state 1 is represents an *pending* request — the first request transition has occurred but has not been acknowledged; state 2 represents a request which has been *granted* by a transition on the acknowledge line (but no further request transitions have occurred); in state 3, the resource has been *released* by a second transition on the request line, but this has not yet been acknowledged. Four-phase interfaces march through these states in this order, or stop forever in one of them. In the even-numbered states, a request transition is the next to occur; in odd states, acknowledges are next.

Finally, a user is said to be in the *critical region* if its interface is in state 2 or 3; the user has the resource and no other request should be granted.

LISP functions can be defined to add some syntactic sugaring for four-phase signaling:

```
(defun quiescentp (u) (= u 0))    ; in quiescent state.
(defun pendingp   (u) (= u 1))    ; in pending state.
(defun grantedp   (u) (= u 2))    ; user has the resource.
(defun releasedp  (u) (= u 3))    ; in released state.
(defun ackokp     (u) (oddp u))   ; acknowledges are allowed.
(defun reqokp     (u) (evenp u))  ; requests are allowed.
(defun criticalp  (u) (> u 1))    ; user is in critical region.
```

It is a LISP convention that the names of *predicates* (functions that return true or false) should end with "P".

It is now possible to write a somewhat readable specification for a mutual exclusion element, as is shown in Figure 5.2.

```
(setq ME
   (SM (ur1 ur2) (ua1 ua2) ((u1 4) (u2 4)) ()
       ;; now, the clauses:
       ((or (and (wire-is ur1) (reqokp u1))
            (and (wire-is ua1)
                 (or (and (pendingp u1) (not (criticalp u2)))
                     (releasedp u1))))
        ;; action for first clause.
        (advance u1))
       ;; second clause
       ((or (and (wire-is ur2) (reqokp u2))
            (and (wire-is ua2)
                 (or (and (pendingp u2) (not (criticalp u1)))
                     (releasedp u2))))
        ;; action for second clause.
        (advance u2))))
```

**Figure 5.2.**   LISP Macro to Define Mutual-Exclusion Elements

The `setq` at the beginning of Figure 5.2 stores the resulting trace structure in the LISP variable ME, for future use. The first list argument to SM, (`ur1` `ur2`), gives the input symbols; the second gives the outputs; the third declares two state variables, $u1$ and $u2$ which represent the states of the four-phase interfaces $(ur1, ua1)$ and $(ur2, ua2)$. If a variable is omitted from the initial values list, it is initialized to 0 by default, so the fourth argument says that both interfaces start out in the quiescent state (0).

The remaining arguments are two clauses that define the next-state function. The actions of the clauses compute the next state as a function of the values of the state variables; in this case, the action is a call to the macro `advance`, which increments the named state variable, modulo its declared range (in this case, 4). This represents a four-phase interface moving to its next state. The predicate in the first clause says that the interface should be advanced whenever the current wire is $ur1$ (the request line for $u1$) and a request is allowed by the interface. The interface can also be advanced if the wire is $ua1$ (the acknowledge for $u1$) and there is either a pending request and no one else is in the critical region, or the resource has been released but this has not yet been acknowledged. The clause for the $u2$ interface is similar, except some of the names have been changed.

As an example, consider the state in which $u1 = u2 = 1$. The first predicate is false when the wire is $a$ because (`reqokp` `u1`) is false — the first transition on the request line has not been acknowledged, so four-phase signaling forbids another request. The first predicate *is* satisfied when the wire is $ua1$, because a request from $u1$ is pending and the other interface does not have the resource. This allows the ME element to grant the resource to $u1$. On the other hand, in the state in which $u1 = 1$ and $u2 = 2$, the $u1$ interface cannot be advanced when the wire is $c$ because the condition (`not` (`criticalp` `u2`)) fails. This is the part of the specification that ensures mutual-exclusion.

### 5.2.4. Expressions

Once a set of primitive SPCTS has been defined, they can be used to construct large circuit descriptions using the operations of circuit algebra, called COMPOSE, HIDE, and RENAME in the LISP implementation. For convenience, COMPOSE takes *one or more* trace structures; this is the usual extension of an associative binary operation to any number of arguments. HIDE takes a list of wire names to be hidden and a subexpression. RENAME takes a list of ordered pairs representing the renaming function and a subexpression.

As we shall see below, it is not necessary to construct a composite state graph in order to check whether a composite circuit conforms to a specification. This is fortunate, since composite state graphs are often quite large. Instead, these operations construct *expression trees.* There is a node type for each operation except RENAME. The leaves of the trees are always SPCTS nodes and the internal nodes are of types COMPOSE or HIDE. A COMPOSE node has *subexpression list*, which is a list of nodes representing the subexpressions that are composed. A HIDE node has a field containing the list of symbols that are hidden and and a subexpression field with another node in it. Each internal node has its own *names*, *I*, and *O* fields. The RENAME operation does not need an explicit internal node – it simply changes the contents of the *names* field of its operand.

The expressions are in a *normal form* similar to that defined in Section 2.6. A normal-form expression is either a SPCTS, a COMPOSE node with two or more subexpressions which are SPCTS, or a HIDE node which hides at least one symbol in a subexpression which is a normal-form tree having a COMPOSE or a SPCTS node at its root. The operations maintain the trees in normal form by applying the laws of circuit algebra: for example, if two trees with HIDE nodes at their roots are composed, the COMPOSE function renames the hidden nodes as necessary (say, to $D$ and $D'$), then recursively composes the subexpressions under the HIDE nodes, then hides $D \cup D'$. Thus, the HIDE node is moved to the top of the expression.

## 5.3.  Conformation Checking

The LISP function CONFORMS-TO-P checks conformation CONFORMS-TO-P takes two arguments: an expression tree representing a circuit implementation and a SPCTS representing a specification. If the first conforms to the second, the function returns normally; otherwise, it reports an error and prints a list of wire names representing the failing trace, to help the user diagnose the error.

CONFORMS-TO-P does not need to build a composite state graph in order to discover a problem; it can check for failures as the state graph is being built, and report them immediately. This is very important, because the state graph for an incorrect circuit often has orders of magnitude more states than the state graph for a correct circuit; the additional states represent random actions by the circuit after a failure has already occurred. Reporting failures before the entire state graph has been constructed greatly enhances the practicality of automatic verification.

Our implementation of CONFORMS-TO-P reduces the problem of testing conformation to the problem of checking whether the composition of the implementation with the mirror of the specification is failure-free, as discussed in the Chapter 4. The verifier searches the *product graph* representing the composite traces of the composition. The search is *depth-first*; states are created only as they are needed.

In more detail, alphabets of the expression tree and the specification are compared to see if they match — an error is reported if they do not. The expression describing the implementation is then converted into a *components list*, which is a list of SPCTS. If the expression is a single SPCTS node, the list consisting of the node is used. If the expression has a COMPOSE node on top, the subexpression list (a list of SPCTS) is used. Finally, if the tree is a HIDE node, the hidden symbols are ignored and its subexpression (a COMPOSE or SPCTS) is converted to the list via one of the previous two steps. This processing is reminiscent of the steps in the proof of Lemma 4.2. The mirror of the specification is computed by exchanging the I and O bitvectors. The mirror of the specification is then placed on the front of the components list.

Each state in the product graph (called a global state) is a vector of states from the SPCTS on the components list; the $i$th element of this vector is a state from the $i$th SPCTS on the list. The initial global state is the vector of the start states of the individual SPCTS. The graph is recursively constructed and searched depth-first. When the search routines is called on a global state, it first checks a hash table to see if the global state has already been visited. If it has, the routine returns immediately. Otherwise, the routine enters the global state in the hash table, constructs vectors for the successors in the global state graph, and recursively processes the successors.

Given a current global state, the global successor states are computed by inspecting the outputs of each component. The components list has the property that there is exactly one component (the *driver*) which has any particular wire (call it $a$) as an output, so there is a global successor state if and only if the driver has a successor on $a$. In case there is a global successor, it is computed from the alphabets and transition functions of the elements of the component list, as follows. Let us consider the $i$th member of the component list, $c_i$. If $c_i$ does not have $a$ in its alphabet, the $i$th member of the global state does not change. If $c_i$ has $a$ as an input and there is a transition defined on $a$, the $i$th element of the global successor is the component successor. If $c_i$ has $a$ as an input, but *no* transition on $a$, there is a failure state in the global successor; the search routine reports the problem and stops. The routine maintains a list of transitions representing the path it has followed from the initial global state to the current state. In the event that a failure is discovered, this path is a failure trace, which is printed the help the user diagnose the problem.

This trace may be quite long — depth-first search usually does not find the shortest path to a problem. Consequently, after reporting the error, the verifier offers to do a breadth-first search to find the shortest failure trace; it is often much easier to understand the failure from the shorter example. Of course, the implementation could have employed breadth-first search from the beginning; depth-first search was chosen initially because it requires less space to keep track of the remaining global states to be searched — in addition to the table of states it has visited, all it needs to remember is the path by which it reached its current state. On the average, breadth-first search needs to keep all the global states at a particular distance from the root; if the global states have more than one child on the average, the number of these grows exponentially with the depth of the tree. Moreover, in the the examples that have been tried, breadth-first search seems to examine a greater number of states to find an error.

If the implementation does not meet the specification, the problem is usually detected after exploring a small fraction of the global states that would have appeared in the composite graph. If the circuit meets the specification, all of the states will be explored. However, even in this case there are savings compared with constructing a SPCTS for the implementation, because the transitions *between* the states do not need to be explicitly represented.

This algorithm is linear in the number of states examined, which is at most the product of the number of states in the individual components. In practice, the critical resource is the space used by the state table. When verifying a difficult circuit, performance begins to suffer because of conflicts in the state table and paging from secondary storage when the state table becomes too large for primary memory. Some time after this performance degradation, the program runs out of paging space and halts. To some extent, these problems can be addressed by using more compact data structures or a bigger computer. However, if a design inherently requires inspecting an exponentially large state graph, constant-factor improvements in space usage will not substantially increase the size of the circuits that can be verified.

## 5.4.  Example 1:  Tree Arbiter

Mutual exclusion is a universal problem in concurrent programming.arbiter,tree This is no less true of asynchronous circuits than of operating systems. In asynchronous hardware, the shared resource might be a communications bus or a storage element (e.g. a multi-port memory). This example and the next are ways of implementing mutual exclusion with standard cells, so a family of circuits to handle various numbers of users can be generated by composing the cells in some regular way.

The approach of this section is to handle many users by building a *tree* of cells. The designs and the errors in them are by the author. However, the idea and the correct design are borrowed directly from Seitz [70]. Figure 5.3a is a block diagram of a single cell; Figure 5.3b shows how several cells can be connected to handle more than two users.

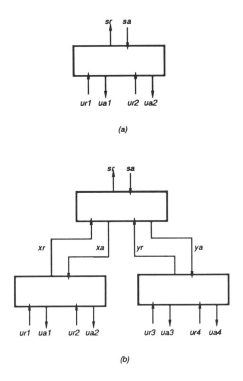

**Figure 5.3.**   Tree Arbiters

Each cell has three four-phase interfaces: $s$ with signals $(sr, sa)$, $u1$ with $(ur1, ua1)$, and $u2$ with $(ur2, ua2)$.

The tree of arbiters is like an elimination tournament. $n$ users are arbitrated two at a time by the individual cells. The $n/2$ winners at each level are arbitrated by the next level, until there is one winner for the entire tree. The requests propagate up to the top of the tree, then the acknowledge signals propagate down the tree (but only to one child at each level) until a user who is the winner of the tournament is acknowledged.

Four-phase signaling is used on all of the request/acknowledge interfaces. When one of the users of an arbiter cell requests the resource, that cell asks its parent for the resource by signaling on the $s$ interface. The transitions on $sr$ and $sa$ have the same meanings as they do at the $u1$ and $u2$ interfaces. If the parent is not at the top of the tree, it will ask *its* parent for the resource, and so on. If the parent grants the resource to the cell, the cell then grants the resource to no more than one of its children. When this child releases the resource to the cell, the cell then releases it to the parent, waits to be acknowledged by the parent, then acknowledges the child.

The arbiter will be verified at two levels of abstraction. At the high level, the cell specification is a description of a primitive in a larger circuit that is a tree of cells. At the low level, the cell specification is a specification to be met by an implementation that uses C-elements, AND gates, and so forth.

### 5.4.1. Specification of an Arbiter Cell

This subsection gives the specification of a single cell in the tree arbiter. The sequence of transitions for one of the $u$ interfaces (say, $u1$) and the $s$ interface should follow the strict sequence

$$ur1; \; sr; \; sa; \; ua1; \; ur1; \; sr; \; sa; \; ua1.$$

During this process, the other user may request the resource at any time, but should not be granted the resource if the first user has it or if the parent does not have it.

The tree arbiter cell is specified using the SM macro. The specification has much in common with the specification of the ME element. There are three state variables called $u1$, $u2$, and $s$, one for each of the interfaces of the cell. Each variable cycles through the values $(0, 1, 2, 3)$, representing the request and acknowledge wire values cycling through $(00, 01, 11, 10)$, just as in the ME description. The state machine is specified by controlling the conditions under which each interface can advance to the next phase.

The SPCTS for the specification is defined using the SM macro, as shown in Figure 5.4.

The inputs to the specification are $ur1$, $ur2$, and $sa$ and the outputs are $ua1$, $ua2$, and $sr$. There are three state variables, $u1$, $u2$, and $s$ in the range $[0 \ldots 3]$. By convention, if a state variable is not mentioned in the initial values list (the third argument), it is 0; in this case, that list is empty, so all of the state variables are 0, initially.

The transition function described by the clauses is somewhat similar to the ME element. The first two clauses are the same as in the ME element, except that there are some extra conditions imposed when the current wire is $ua1$ or $ua2$ to make sure transitions are properly ordered with respect to the $s$ interface. In the predicate of the first clause, the cell is allowed to acknowledge $u1$ if there is a request pending, if $u2$ does not already have the resource, *and* if the resource has been granted to the cell by the parent — the user should not get the resource unless the cell has it to give. There can also be an acknowledge if the user has released the resource, and the $s$ interface has returned to quiescence (by releasing the resource to the parent and having that release acknowledged)

```
(setq TREECELL
      (SM (url ur2 sa) (ual ua2 sr) ((ul 4) (u2 4) (s 4)) ()
          ;; clause for ul interface.
          ((or (and (wire-is url) (reqokp ul))
               (and (wire-is ual)
                    (or (and (pendingp ul)
                             (not (criticalp u2))
                             (grantedp s))
                        (and (releasedp ul) (quiescentp s)))))
           (advance ul))
          ;; clause for u2 interface.
          ((or (and (wire-is ur2) (reqokp u2))
               (and (wire-is ua2)
                    (or (and (pendingp u2)
                             (not (criticalp ul))
                             (grantedp s))
                        (and (releasedp u2) (quiescentp s)))))
           (advance u2))
          ;; clause for s interface.
          ((or (and (wire-is sr)
                    (or (and (quiescentp s)
                             (or (pendingp ul) (pendingp u2)))
                        (and (grantedp s)
                             (or (releasedp ul)
                                 (releasedp u2)))))
               (and (wire-is sa) (ackokp s)))
           (advance s))))
```

**Figure 5.4.**  LISP Specification for a Tree Arbiter Cell

— so the user release is not acknowledged until the parent acknowledges the cell. The *u2* interface is symmetric with this.

The third clause controls the *s* interface. When the current wire is *sr*, the *s* interface should be advanced if it is quiescent and if *u1* or *u2* has requested the resource; this asks the cell's parent for the resource. There should also be a transition on *sr* if the cell has the resource [(grantedp s)], and the user who wanted it has released it. The last part of the predicate says that the interface can receive an acknowledge on *sa* whenever the four-phase protocol allows it.

### 5.4.2. High-level Verification

The tree arbiter cell is used to build bigger arbiters. In this subsection, we use the specification of the cell as a *description of a primitive* in two small trees of cells, which should conform to a specification of the bigger arbiter. This is a good way to make sure that the idea behind the tree arbiter is sound, and more pragmatically, to debug the specification just described.

The first example is a single cell used to implement a two-way ME element (which was specified in the previous section). For example, a single cell with a non-inverting buffer connecting *sr* to *sa* should conform to the specification for the ME element given in Figure 5.2 above. CONFORMS-TO-P confirms that this is the case, examining 28 states in the process.

```
(setq ME3
      (SM (ur1 ur2 ur3) (ua1 ua2 ua3) ((u1 4) (u2 4) (u3 4)) ()
          ;; now, the clauses:
          ((or (and (wire-is ur1) (reqokp u1))
               (and (wire-is ua1)
                    (or (and (pendingp u1)
                             (not (criticalp u2))
                             (not (criticalp u3)))
                        (releasedp u1))))
           (advance u1))
          ;; second clause
          ((or (and (wire-is ur2) (reqokp u2))
               (and (wire-is ua2)
                    (or (and (pendingp u2)
                             (not (criticalp u1))
                             (not (criticalp u3)))
                        (releasedp u2))))
           (advance u2))
          ;; third clause
          ((or (and (wire-is ur3) (reqokp u3))
               (and (wire-is ua3)
                    (or (and (pendingp u3)
                             (not (criticalp u1))
                             (not (criticalp u2)))
                        (releasedp u3))))
           (advance u3))))
```

**Figure 5.5.** LISP Specification for 3-Way Mutual Exclusion

What about three-way mutual-exclusion? The definition of the two-way ME-element can be extended to a three-way element by adding a state variable *u3*, adding an additional clause to check it, and modifying the predicates to check whether *either* of the other two interfaces has is in the critical section before granting to the current user. A definition using SM is shown in Figure 5.5.

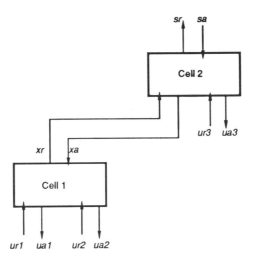

**Figure 5.6.**   Three-way Tree Arbiter

The program was used to check whether the circuit depicted in Figure 5.6 conforms to the three-way ME specification. Surprisingly, CONFORMS-TO-P reported an error. The counterexample trace from the initial depth-first search is 73 transitions long. However, the counterexample from breadth-first search has 15 transitions:

$$ur1; ur3; xr; sr; sa; xa; ua1; ur1; xr; sr; sa; xa; sr; sa; ua3.$$

Here is what happened: *u1* requested the resource, and so did *u3* (*ur1; ur3*); the sequence *xr; sr; sa; xa* is *cell1* asking for the resource and getting it; then *u1* releases, and so do both cells (*ur1; xr; sr; sa; xa*); then *cell2*, which still has a pending request from *u3*, asks for the resource again and gets it (*sr; sa*); and then it grants to *u3* (*ua3*). This is seems reasonable, except that *ua1* is still in the critical section when the resource was granted to *u3*, which violates the specification of the three-way ME element.

The specification of the cell could perhaps be modified to conform to the specification. However, a better solution is to change the specification of a three-way ME element (in this case the bug is in the specification, not the implementation). The current specifications are based on the idea that the critical section consists of phases 2 and 3. However, mutual exclusion in phase 3 is neither necessary nor useful. Once a resource has been released, the user should be through with it; at that point, it should be available to other users. The critical section should consist only of phase 2. The specification should only say that no two users can be in phase 2 at the same time. This is repaired easily by redefining `criticalp` to be the same as `grantedp`. (In the experiments here, only the definition of the three-way ME element was changed, although the definition of `criticalp` in the other specifications could be changed also without ill effect.)

The tree of two cells conforms to this specification (CONFORMS-TO-P examines 169 states to check this). Although there were no bugs in the implementation at this level, we have better grounds for believing that the specification is correct, and we have corrected an misconception about critical sections in the specifications of mutual-exclusion elements.

### 5.4.3. Low-level Verification

Figure 5.7a represents a first, rather naive, attempt to implement the arbiter cell. The OR is used to raise *sr* whenever either user has requested the resource, and the AND gates are used to steer the acknowledge from the parent (*sa*) to the user whose request was granted. CONFORMS-TO-P complains about the trace

$$ur1; ur2; g1; sr; sa; ua1; ur1; g1; g2$$

in the depth-first phase. The problem is that when *g1* falls, the mutual-exclusion element can immediately raise *g2* in response to the request from *ur2* — but this causes a hazard in the OR gate, which has not had a time to respond to the change in *g1*. The breadth-first phase finds a different problem:

$$ur1; g1; sr; sa; ua1; ur1; g1; ua1.$$

When *g1* falls, the upper AND gate may respond by lowering *ua1*; but this is premature, since the specification requires that the *s* interface return to 0 first. The pitfall in this design was not thinking about the falling transitions.

An improved circuit appears in Figure 5.7b. The additional ANDs with one inverted input are used to prevent the rest of the circuit from seeing changes in *g1* and *g2* until the acknowledge for the other interface has returned to zero. The two ANDs driving

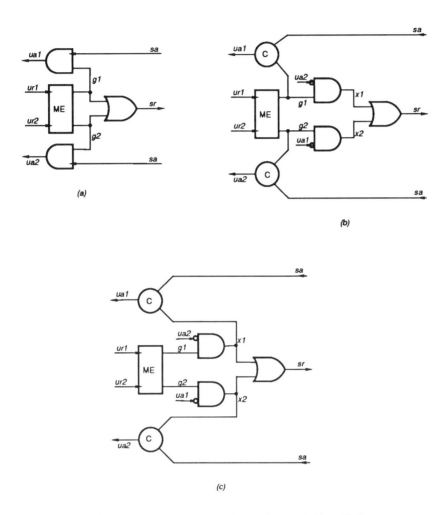

**Figure 5.7.** Implementations of Tree Arbiter Cell

the user-acknowledge wires have been changed to C elements (Section 3.4), so that the acknowledges will not return to zero until both *sa* and *g1* (or *g2*) have returned to zero. CONFORMS-TO-P complains about the trace

$$ur1; ur2; g1; x1; sr; sa; ua1; ur1; g1; g2; ua2$$

(in the breadth-first phase). An acknowledge on *ua1* can occur before the *s* interface has returned to zero. The problem is that after *g1* goes low, *g2* can go high before *sa* goes low, causing the lower C element to raise *ua2* before it should.

The design can be fixed by using the *x* signals instead of the *g* signals as inputs to the C elements. The *x* signals follow the *g* signals, but they are delayed until the *s* interface has returned to zero. Figure 5.7c shows the amended circuit. CONFORMS-TO-P reports that this circuit meets the specification; 52 states are examined in discovering this.

## 5.5.  Example 2: Martin's Distributed Mutual Exclusion Circuit

Martin has proposed a different solution to the *n*-way mutual exclusion problem [49]. In his solution, the arbiter cells are arranged in a ring. Each is connected to a user and to identical cells on the left and right. There is a single *token* that inhabits no more than one cell in the ring at any time. A cell may receive requests for the token from its user or from the cell on its left. If the cell has the token, it gives it to whomever requested it (if both request the token, the cell arbitrarily decides between them). If the cell does not have the token but has a request for it, it requests the token from the cell to the right. When the user releases the resource, the token reappears in the corresponding cell. Hence, in a typical case, a user requests the resource, causing requests for the token to propagate around the ring to the right until they reach the cell that has the token; then the token propagates to the left until it reaches the cell whose user made the request. The token then disappears until the user has released the resource, at which point another cycle begins. Unlike the tree arbiter, this arbiter is completely symmetric.

This example is interesting for several reasons. First, it is one of the larger fully speed-independent circuits published. Second, the verifier discovered a problem in the circuit even though the circuit was derived by transformations from a correct program, demonstrating that automatic verification is useful even when a circuit has been designed carefully.

### 5.5.1. Specification of the DME cell

As before, we begin by specifying the trace structure for the cell. A block diagram of the cell appears in Figure 5.8. There are three four-phase interfaces: $u$ with wires $(ur, ua)$, $r$ with $(rr, ra)$, and $l$ with $(lr, la)$. The $u$ interface receives requests for the resource from a user, the $l$ interface receives requests from a cell to the left, and the $r$ interface makes requests on a cell on the right.

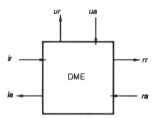

**Figure 5.8.** Distributed Mutual Exclusion Cell

The circuit was specified in the original paper using a CSP-like language. Unfortunately, the CSP specification cannot be translated directly into a trace structure because it does not completely specify the implementations of the communication protocols at the various interfaces. Instead, the SM macro is used. The specification appears in Figure 5.9*. The inputs are $lr$, $ur$, and $ra$, and the outputs are $la$, $ua$, and $rr$. There are four state variables: one for each of the four-phase interfaces ($l$, $u$, and $r$), and one representing whether the cell has the token: $b = 1$ if it has the token, and $b = 0$ if it does not. The initial values of all the state variables are 0 except for $b = 1$.

There are three clauses to control each four-phase interface. In addition to cycling through the four-phase protocols, these clauses sometimes change the $b$ variable. For the $l$ interface, the rules say that an $lr$ transition can occur whenever a request is allowed by the four-phase protocol. An outstanding request on $l$ is acknowledged only if the cell has the token ($b = 1$); in this case, the $l$ interface advances to the next phase, and the value of $b$ is set to 0 to reflect the fact that the cell has given the token to its left neighbor. The

---

*Steven Nowick has pointed out that this can generate redundant requests on the right interface in some situations. This phenomenon appears to do no harm, but it is inconsistent with our intention in writing the specification

```
(setq DMECELL-TOK
  (SM (lr ur ra) (la ua rr) ((l 4) (u 4) (r 4) (b 2)) ((b 1))
      ((and (wire-is lr) (reqokp l))
       (advance l))
      ((and (wire-is la) (pendingp l) (= b 1))
       (advance l)
       (setq b 0))
      ((and (wire-is la) (releasedp l))
       (advance l))
      ((and (wire-is ur) (reqokp u))
       (advance u))
      ((and (wire-is ua) (pendingp u) (= b 1))
       (advance u)
       (setq b 0))
      ((and (wire-is ua) (releasedp u))
       (advance u)
       (setq b 1))
      ((and (wire-is rr)
            (or (and (quiescentp r)
                     (or (pendingp l) (pendingp u)) (= b 0))
                (grantedp r)))
       (advance r))
      ((and (wire-is ra) (pendingp r))
       (setq b 1)
       (advance r))
      ((and (wire-is ra) (releasedp r))
       (advance r))))
```

**Figure 5.9.**   LISP Specification for DME Cell


*la* wire is also allowed to change to acknowledge a release on the *l* interface (without changing the status of the token).

The description of the *u* interface is similar. The only difference is that *b* is set to 1 (instead of 0) when the user acknowledge returns to zero; in essence, the user returns the token to the cell when it is done.

For the *r* interface, a request is sent out on *rr* whenever the interface is quiescent ($r = 0$) and the cell wants the token because there is an outstanding request from the left neighbor or the user, and the cell does not already have the token. *rr* may also change when the cell has received the token and stored it. If there is a pending request on *r*, the request is acknowledged only if the cell's right neighbor has given it the token. In this case, the next state has $b = 1$ in addition to advancing the $r$ interface, the cell has acquired the token. Finally, $ra$ may return to zero when $r$ has been released.

The state machine may start in one of two states, depending on whether it has the token ($b = 0$) or not ($b = 1$). The specification above shows the second case. In either case, all of the other variables start at 0. There is only one change that needs to be made in the description to describe a cell starting in the state in which it does not have the token: $b$ is initialized to 0 instead of 1.

### 5.5.2. High-level Verification

Given the complexity of this specification, it would be wise to check it by verifying that small multi-way arbiters built from the cells meet their specifications. Several typographical errors in earlier versions of the specification were uncovered in the specification by this method. Once again, the specification of the cell is used as a *description of a primitive* at this level of abstraction.

The first test is to see whether a single cell in which the $l$ interface has been connected to the $r$ interface via two non-inverting buffers (Figure 5.10a) conforms to a "one-way mutual-exclusion element", i.e., a buffer. CONFORMS-TO-P confirms that this is the case by examining four states. The specification of the cell must start in the state in which it has the token; otherwise, the cell meets the specification by deadlocking.

The next test is to see if two cells, connected as in Figure 5.10b, conform to to a two-way mutual exclusion element. Initially, the left cell has the token and the right does not. The specification passes the test; 208 states are inspected in checking this.

The test with three cells also passes (after checking 2,496 states). Interestingly, this system conforms to the original specification of a three-way mutual exclusion element (with the critical section defined as phases 2 and 3), which the tree arbiter failed to implement.

### 5.5.3. Low-level Verification

Martin's published implementation of the DME cell appears in Figure 5.11. Several of the components are somewhat different from previous descriptions of the same components in Section 4.5. One of these is the SR latch. The SM description for the version of the latch used in this chapter is shown in Figure 5.12.

In English, the inputs $s$ and $r$ are allowed to change whenever both are low, or when one is high and the corresponding output ($b$ for $s$ and $notb$ for $r$) is also high. The output $b$ can change when $r = 1$ and when $b = 1$ (it goes low because the latch is being reset) or when $s = 1$ and $notb = 0$ (it is being set and the other output has already gone low). The conditions for $notb$ are similar. The difference between this description and the description of Section 4.5 is that when the latch changes state, the outputs are guaranteed

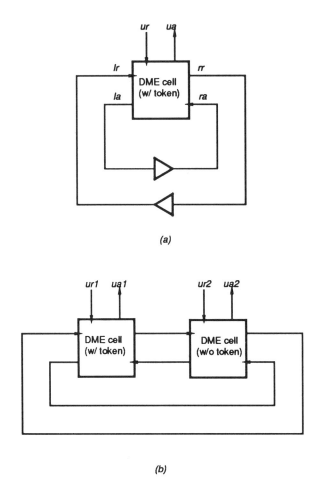

**Figure 5.10.** One- and Two-Way Distributed Mutual Exclusion

to change in a fixed order because both must be 0 at some time. This restriction makes the circuit more convenient to use in speed-independent circuits: the environment can tell that the latch has definitely settled in the "set" state when $b = 1$, and that it is in the "reset" state when $b = 0$.

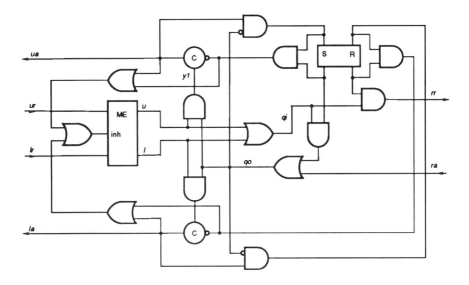

**Figure 5.11.**   Published Implementation of DME Cell

The mutual-exclusion element in this circuit is slightly different from the previous one because it has an additional signal, *inh*. If *inh* is high, the ME element waits for it to go low before it grants to one of the users. The description of the mutual-exclusion element is modified by adding this input, that is allowed to change at any time. The description is similar to that of the ME element except that there is an additional condition: *inh* = 1 must be true for either user to enter the critical region. The modified description appears in Figure 5.13.

To specify the initial conditions for the circuit, we assume that neither the implementation nor the specification has the token, initially. In the implementation, this means that the SR latch has *b* = 0 and *notb* = *1*. All other signals are initially 0. When CONFORMS-TO-P is asked whether the implementation conforms to the specification, it examines 12 states, then reports the error

$$lr; ur; u; qi; rr; ra; qo; y1; ua; ur; u; l.$$

This path is produced by both depth- and breadth-first search. The immediate problem is that there is a hazard in the OR gate which has *l* and *u* as inputs. When *u* goes low, there is an outstanding request on *lr* which the ME element can grant immediately — the

```
(setq SR-FF
      (SM (s r) (b notb) ((s 2) (r 2) (b 2) (notb 2)) ((notb 1))
         ((and (wire-is s)
               (or (and (= s 0) (not (= r 1)))
                   (and (= s 1) (= b 1))))
          (advance s))
         ((and (wire-is r)
               (or (and (= r 0) (not (= s 1)))
                   (and (= r 1) (= notb 1))))
          (advance r))
         ((and (wire-is b)
               (or (and (= r 1) (= b 1))
                   (and (= s 1) (= notb 0) (= b 0))))
          (advance b))
         ((and (wire-is notb)
               (or (and (= s 1) (= notb 1))
                   (and (= r 1) (= b 0) (= notb 0))))
          (advance notb)))))
```

**Figure 5.12.** LISP Specification for SR Latch

inhibit signal which was supposed to prevent this can be delayed in various OR gates, and may not arrived at the ME element in time. The more general problem is that the delays in the inhibit logic were not taken into account in the design of the circuit. The transformations that were used to derive the rest of the circuit were not used for this part of it, resulting in the introduction of a bug.

Martin also discovered this problem after publishing the circuit, and redesigned the circuit. The modified circuit appears in Figure 5.14. An ordinary ME element is used in this circuit; external circuitry is used to delay the processing of one request until the processing of the other has been completed. Interestingly, the left part of the circuit is almost identical to the final implementation of the tree arbiter. This implementation *does* meet the specification; CONFORMS-TO-P examines 154 states to discover this.

```
(setq ME-inh
      (SM (ur1 ur2 inh) (ua1 ua2) ((u1 4) (u2 4) (inh 2)) ()
          ((or (and (wire-is ur1) (evenp u1))
               (and (wire-is ua1)
                    (or (and (pendingp u1)
                             (not (criticalp u2))
                             (not (= inh 1)))
                        (releasedp u1))))
           (advance u1))
          ((or (and (wire-is ur2) (reqokp u2))
               (and (wire-is ua2)
                    (or (and (pendingp u2)
                             (not (criticalp u1))
                             (not (= inh 1)))
                        (releasedp u2))))
           (advance u2))
          ((wire-is inh)
           (advance inh)))))
```

**Figure 5.13.**   LISP Macro for Mutual Exclusion Element with Inhibit

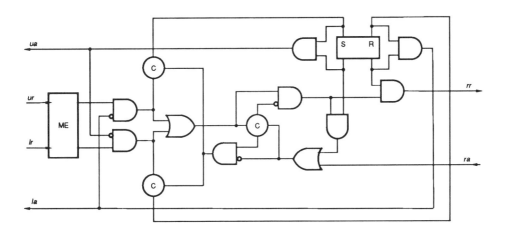

**Figure 5.14.**   A Correct Implementation of the DME Cell

## 5.6.  A Trace Theory Calculator

To do the examples of the previous two sections, it is not necessary to *evaluate* expressions of circuit algebra, that is, to translate the expression into a single simple prefix-closed trace structure. However, in some circumstances evaluating the expression may be appropriate, anyway. First, it may be important for efficiency. If a particular sub-circuit appears a number of times in an implementation, evaluating and then simplifying may reduce the size of the state graph that needs to be explored. Second, determining some interesting properties of circuits requires evaluation of expressions. For example: what is the least *delay-insensitive* specification to which a composite circuit conforms? To find out, the **DI** operator (written in LISP) can be applied the resulting expression evaluated. Also, determining whether two composite circuits are conformation-equivalent requires that both be evaluated, CONFORMS-TO requires the specification to be a SPCTS. Finally, inspecting the state graph for a composite circuit can sometimes yield insights into the operation of the circuit.

The program has a function called TEVAL which evaluates a circuit algebra expression. If the expression is already a SPCTS, TEVAL returns it. There are no RENAME expressions, since RENAME applied to an expression changes the names directly. The evaluation of the remaining expression types, COMPOSE and HIDE, is more involved.

If the expression is of type COMPOSE, TEVAL constructs a global state graph using an algorithm similar to CONFORMS-TO-P. However, the algorithm is somewhat more complicated because it must keep track of the transitions between the states, as well as the states themselves. The resulting state graph contains autofailure states which must be removed.  This is done by a depth-first traversal of the *reverse* of the state graph, starting from the states in the list of autofailures. Any predecessor of an autofailure state which reaches the autofailure via an output symbol is recursively marked as an autofailure. When this procedure terminates, all of the states which have been marked as autofailures are removed from the state graph. This procedure takes care of all the simplifications of the previous chapter: the recursive marking procedure is an implementation of autofailure manifestation, and removing the states from the state graph takes care of failure exclusion,implementation of.

TEVAL of a HIDE node builds a non-deterministic finite state automaton by converting all transitions on hidden symbols to transitions on $\epsilon$ [44]. This automaton is then converted to a deterministic automaton using the subset construction, which is slightly modified to keep track of failure states. If a set of states in the subset construction contains a failure, the state is removed. This takes care of the failure-exclusion transformation the previous chapter. Autofailures are not introduced by hiding, so they do not need to be removed.

# Chapter 6

# Infinite Sequences and Infinite Games

## 6.1. Introduction

Liveness properties are assertions about the behavior of a system into the indefinite future. Consequently, Chapter 7, which develops a trace theory for liveness properties, makes heavy use of *infinite* sequences. This chapter lays a foundation for the next by presenting mathematical background on infinite sequences and some results on infinite games.

Many of the definitions and results for finite sequences apply also to infinite sequences; for example, there are regular languages of infinite sequences and finite automata that accept them. The second and third sections cover definitions and basic results for infinite sequences and finite automata on infinite sequences.

The definitions of receptiveness and autofailure for the trace theory of Chapter 7 are most naturally defined using *infinite games*, which are two-player games that are played forever. This material is likely to be unfamiliar. The fourth section defines infinite games and presents some basic results. The fifth section proves a new result: for a particular class of infinite games, the problem of whether the second player has a winning strategy is decidable. This is a crucial result, because it is the basis for a decision procedure for receptiveness and for autofailure manifestation for the trace structures of Chapter 7.

## 6.2. Mathematical Preliminaries: Infinite Sequences

This section presents definitions and basic results relating to infinite sequences.

An $\omega$-*sequence over* $X$ (or an *infinite sequence*) is a function from $\omega$ to $X$. The set of all $\omega$-sequences of type $\omega \to A$ is written $A^\omega$. The length of an $\omega$ sequence $x$ ($|x|$) is $\omega$. $i \leq \omega$ means either $i \in \omega$ or $i = \omega$. Without modification, the definition of concatenation of Section 3.2 specifies $xy$ when $x$ is a finite sequence and $y$ is an $\omega$-sequence. This concatenation is not defined when $x$ is infinite. The concatenation of sets $XY$ is defined when all of the sequences in $X$ are finite. The definition of a prefix is also the same as in Section 3.2. When $z$ is an $\omega$-sequence, we define **pref**$(z)$ to be the set of all *finite* prefixes of $z$. **pref** is extended to sets of sequences as in Section 3.2. The set of all finite and infinite sequences over $A$ is written $A^\infty$ $(= A^* \cup A^\omega)$.

A set $X$ is a *prefix chain* if it is totally ordered by the prefix ordering. The *limit* of a prefix chain $X$, written $\lim(X)$, is the least sequence such that for every $x \in X$, $x \leq \lim(X)$ (note that the definition of $\leq$ for sequences from Section 3.2 applies to infinite sequences). If $X$ has a maximum element, that element is $\lim(X)$ and if $\lim(X)$ is finite, then $\lim(X) \in X$. If $Y$ is a set of sequences and $X \subseteq Y$, then $X$ is a *maximal chain* (in $Y$) if for every prefix chain $X'$, $X \subseteq X' \subseteq Y$ implies $X' = X$. $\lim(Y)$ is defined to be the set $\{\lim(X) \mid X$ is a maximal chain in $Y\}$. For every $z \in A^\infty$, **pref**$(z)$ is a prefix chain. It is easy to show that

6.1                                                   $\lim[\textbf{pref}(z)] = z.$

The following two lemmas are very useful in proving the equivalence of infinite sequences:

**Lemma 6.1.** *Given two prefix chains $X$ and $Y$, if for every $x \in X$ there is a sequence $y \in Y$ such that $x \leq y$, then $\lim(X) \leq \lim(Y)$.*

**proof.** If $y \in Y$ then $y \leq \lim(Y)$. Since for every $x \in X$ there is a $y$ such that $x \leq y$, then $\forall x : x \leq \lim(Y)$. But then by the definition of lim, $\lim(X) \leq \lim(Y)$. □

**Lemma 6.2.** *If $X$ and $Y$ are prefix chains of finite sequences and $\lim(X) \leq \lim(Y)$, then for every $x \in X$ there is a $y \in Y$ such that $x \leq y$.*

**proof.** Since $\lim(X) \leq \lim(Y)$, for every $x \in X$ and $y \in Y$, $x \leq \lim(X) \leq \lim(Y)$ and $y \leq \lim(Y)$, so $x$ and $y$ are in **pref**$[\lim(Y)]$, a chain, so they are totally ordered: either $x \leq y$ or $y \leq x$. Suppose that for some $x \in X$ there is no $y \in Y$ such that $x \leq y$. Then for every $y \in Y$, $y < x$, so then $y < \lim(X)$, so $\lim(Y) \leq \lim(X)$. By hypothesis, $x$ is finite so $\lim(Y)$ is finite, so $\lim(Y) \in Y$. But then $\lim(Y) < x$, also. Hence $\lim(Y) < \lim(X)$, which contradicts the assumption that $\lim(X) \leq \lim(Y)$. So for every $x \in X$ there is a $y \in Y$ such that $x \leq y$. □

Concatenation can be generalized to sequences of sequences. First, it is extended to finite sequences. Let $\sigma$ be a finite sequence of sequences ($\sigma : i \rightarrow A^*$, where $i \in \omega$) The concatenation of $\sigma$ is defined recursively on $i$: if $i = 1$, it is $\sigma(0)$; otherwise, it is the concatenation of $\sigma(0)$ with the concatenation of the sequence $\sigma(1), \ldots, \sigma(i-1)$. The concatenation of an $\omega$-sequence $\sigma$ is defined to be the limit of the concatenations of its prefixes.

Concatenation can also be extended to sequences of sets of finite sequences (of type $i \to 2^{A^*}$). A sequence of sequences $\sigma$ is *elementwise contained* in a sequence of sets of sequences $\Sigma$ if for all $i \in \omega$, $\sigma(i) \in \Sigma(i)$. The concatenation of $\Sigma$ is defined to be the set of the concatenations of the sequences that are elementwise contained in $\Sigma$.

Some notational devices are needed to distinguish sequences from sequences of sequences. If $A$ is a set not containing sequences and $x_i \in A^*$ for all $i \in \omega$, we write sequences of the $x_i$ with commas separating the individual sequence elements: e.g. $x_0, x_1, \ldots, x_{i-1}$ for a finite sequence of length $i$ and $x_0, x_1, \ldots$ for an $\omega$-sequence. The *concatenations* of these sequences (members of $A^\infty$) are written $x_0 x_1 \ldots x_i$ and $x_0 x_1 \ldots$. The same convention is used to distinguish sets of sequences from sequences of sets: if $X_i \subseteq A^*$ for all $i \in \omega$, then $X_0, X_1, \ldots$ is the sequence of $X_i$'s, while $X_0 X_1 \ldots$ is the concatenation of this sequence, a set of sequences.

If $X$ is a subset of $A^*$, the set of finite sequences having elements in $X$ is written $X^{(*)}$, the set of $\omega$-sequences is $X^{(\omega)}$, and the union of these sets is $X^{(\infty)}$. Note that $X^{(*)}$, and so on, are sets of sequences of sequences. The sets of sequences $X^*$, $X^\omega$, and $X^\infty$ can be defined as the concatenations of $X^{(*)}$, $X^{(\omega)}$, and $X^{(\infty)}$, respectively.

A function $f$ from finite sequences to finite sequences *respects prefixes* if whenever $x \leq y$, $f(x) \leq f(y)$. If $X$ is a prefix chain and $f$ respects prefixes, $f(X)$ is also a prefix chain. Such a function can be extended to infinite sequences by setting $f(z) = \lim(f[\mathbf{pref}(z)])$.

**Lemma 6.3.** *If $f \in [A^\infty \to A'^\infty]$ and $f(z) = \lim(f[\mathbf{pref}(z)])$ for all $z \in A^\infty$, then $f$ respects prefixes.*

   **proof.** Let $x$ and $y$ be any members of $A^\infty$. If $x \leq y$, then $\mathbf{pref}(x) \subseteq \mathbf{pref}(y)$, so $f[\mathbf{pref}(x)] \subseteq f[\mathbf{pref}(y)]$. But then by Lemma 6.1, $\lim(f[\mathbf{pref}(x)]) \leq \lim(f[\mathbf{pref}(y)])$, so $f(x) \leq f(y)$.   $\square$

The next lemma is a useful fact about such functions.

**Lemma 6.4.** *If $f$ is a function from $A^\infty$ to $A'^\infty$ such that for all $x \in A^\infty$, $f(x) = \lim(f[\mathbf{pref}(x)])$, and if $X$ is a prefix chain in $A^*$, then $f[\lim(X)] = \lim[f(X)]$.*

   **proof.** Let $x = \lim(X)$. Since $x = \lim[\mathbf{pref}(x)]$, $x = \lim(f[\mathbf{pref}(x)])$. By Lemma 6.2, for every $y \in \mathbf{pref}(x)$ there is a $y' \in X$ such that $y \leq y'$, and for every $y' \in X$ there is $y \in \mathbf{pref}(X)$ such that $y' \leq y$. Since $f$ respects prefixes by the previous lemma, the same relationship holds between $f[\mathbf{pref}(x)]$ and $f(X)$; so by Lemma 6.1, $\lim(f[\mathbf{pref}(x)]) = \lim[f(X)]$. Hence $f[\lim(X)] = \lim[f(X)]$.   $\square$

Any function that has been naturally extended to sequences respects prefixes, including
**del**($D$) (for any $D$) and renaming functions **r**. A further interesting property of **del**($D$)
is that even if $z$ is an infinite sequence, **del**($D$)($z$) may be finite: for example, if $z = a^\omega$,
**del**($\{a\}$)($z$) = $\epsilon$.

## 6.3.  Regular Languages of Infinite Sequences

There is a theory of regular languages on infinite sequences, and there are finite-state
automata that accept these languages. Most of the results for regular languages of finite
sequences carry over to infinite sequences, as well, although the theory is somewhat
more complicated. Regular sets of infinite sequences are called $\omega$-*regular sets* [24,32].
$\omega$-regular sets are closed under union, intersection, and complementation; furthermore,
if $X$ is regular and $Y$ is $\omega$-regular, $XY$ is $\omega$-regular, too.

The $\omega$-regular sets are defined as the sets described by $\omega$-*regular expressions*, which are
generated by adding an $\omega$ operator to regular expressions. If $\alpha$ is a regular expression and
$\beta$ is regular expression whose set does not contain $\epsilon$, $\alpha \cdot (\beta^\omega)$ is an $\omega$-regular expression;
if $\gamma_1$ and $\gamma_2$ are both $\omega$-regular expressions, so is $(\gamma_1 + \gamma_2)$. The set of $\omega$-sequences
$[\alpha \cdot (\beta^\omega)]$ is defined to be $[\alpha][\beta]^\omega$, and $[\gamma_1 + \gamma_2]$ is $[\gamma_1] \cup [\gamma_2]$. A set is $\omega$-regular if and
only if it can be written as an $\omega$-regular expression.

Like regular sets, there is a form of finite automaton that accepts $\omega$-regular sets. A
deterministic or nondeterministic *Büchi automaton* has exactly the same structure as a
finite automaton, except that we call the set of "final states" $Q_B$ instead of $Q_F$. The
difference between a Büchi automaton and a finite automaton is the way it accepts a
sequence. If $p$ is a *run* of a nondeterministic Büchi automaton on an $\omega$-sequence $x$,
it is a sequence in $Q^\omega$ such that $p(0) \in Q_0$ and for all $i \in \omega$, $p(i+1) \in \mathbf{n}[p(i), x(i)]$
(for deterministic Büchi automata, $p(i+1) = \mathbf{n}[p(i), x(i)]$). For any infinite sequence
$x$, we define In($x$) to be the set of elements that appear infinitely often in $x$; formally,
In($x$) = $\{a \mid x^{-1}(a)$ is infinite$\}$. $p$ is an *accepting run* if In($p$) $\cap Q_B \neq \emptyset$ (at least one state
in $Q_B$ is repeated infinitely often). An automaton accepts $x$ if there is some accepting
run of the automaton on $x$.

We are concerned almost exclusively with nondeterministic Büchi automata. Given
two nondeterministic Büchi automata that have the same alphabet $A$ and that accept
$\omega$-regular sets $X$ and $Y$, there are effective procedures for finding automata that accept
$X \cup Y$, $X \cap Y$ [64], and $A^\omega - X$ (the complement of $X$) [71]. If the automata have $m$ and
$n$ states, the resulting automata have no more than $m + n$, $3mn$, and $16^{n^2}$ states, respectively. Complementation of nondeterministic Büchi automata is PSPACE-complete, so it
is unlikely that a subexponential algorithm will be found in the near future.

Given a Büchi automaton $\mathcal{M}$ accepting the set $X$, an automaton $\mathcal{M}'$ can be constructed that accepts $\mathbf{del}(D)^{-1}(X)$ (where $D \cap A = \emptyset$). The alphabet $A'$ of $\mathcal{M}'$ is $A \cup D$. For each state $q_i$ in $Q_B$, we choose a new state $\hat{q}_i \notin Q$; call the set of all the new states $\hat{Q}$. We set $Q' = Q \cup \hat{Q}$ and $Q'_B = Q_B$. $\mathbf{n}' : Q' \times A' \to 2^{Q'}$ is defined so that

$$\mathbf{n}'(q_i, a) = \mathbf{n}(q_i, a) \quad \text{when } q_i \in Q \cup \hat{Q} \text{ and } a \in A,$$

$$\mathbf{n}'(q_i, a) = \{q_i\} \quad \text{when } q_i \in (Q \cup \hat{Q}) - Q_B \text{ and } a \in D, \text{ and}$$

$$\mathbf{n}'(q_i, a) = \{\hat{q}_i\} \quad \text{when } q_i \in Q_B \text{ and } a \in D.$$

In a diagram of the machine, this construction would introduce a self-loop on $D$ for all non-accepting states. This cannot be done for the accepting states, because it can incorrectly introduce strings ending in $D^\omega$ into the language. So for the accepting states, $D$ symbols go to a non-accepting copy of the state. Any symbol from $A$ will take the machine back from one of the copied states to one of the original states. Note that $\mathcal{M}'$ has $|Q| + |Q_B|$ states.

There is one way in which the theory of $\omega$-regular sets is markedly different from the theory of regular sets: deterministic Büchi automata are strictly less powerful that nondeterministic Büchi automata. For example, the set $(a+b)^* b^\omega$ is not accepted by any deterministic Büchi automaton, but it is accepted by a nondeterministic automaton with two states. The sets accepted by deterministic Büchi automata are closed under union and intersection, but not under complementation or deletion.

Most of the sets of sequences discussed in this chapter contain both finite and infinite sequences. Given a set $X \subseteq A^\infty$, the set of all finite sequences in $X$ is $X \cap A^*$, and the set of all $\omega$-sequences is $X \cap A^\omega$. $X \subseteq A^\infty$ is a *mixed regular* set (or just *regular* when no confusion should result) when $X \cap A^*$ is regular and $X \cap A^\omega$ is $\omega$-regular. Mixed regular sets are closed under union, intersection, complementation, deletion, and inverse deletion. Additionally, if $X$ is regular and $Y$ is mixed-regular, $XY$ is mixed-regular.

Regular expressions can be extended to describe mixed regular sets by allowing $\alpha + \beta$ to be an expression if $\alpha$ is a regular expression and $\beta$ is an $\omega$-regular expression (but, in this case, $\alpha + \beta$ may not be embedded in another regular or $\omega$-regular expression). These are called *mixed-regular* expressions.

*Mixed* finite automata, which accept mixed regular languages, can be defined. A mixed automaton is a finite automaton $\mathcal{M}$ with two sets of accepting states: $Q_F$ and $Q_B$. The states of $Q_F$ are the *finitely accepting* states and those of $Q_B$ are the *infinitely accepting* states. The language of a mixed automaton is the union of the language accepted by the finite automaton with accepting set $Q_F$ and the language accepted by the Büchi automaton with accepting set $Q_B$. Union, intersection, complementation, deletion, and inverse deletion can all be performed on mixed automata at no greater expense than on Büchi automata.

To prove that mixed regular languages are closed under deletion, it is helpful to extend nondeterministic mixed automata to allow transitions on $\epsilon$ in addition to transitions on members of $A$. Informally, a run of the automaton can make any finite number of $\epsilon$ moves between reading symbols from the input, but it must eventually read the entire input. Note that in such an automaton, there can be infinite runs on finite inputs, in case the run has an infinite number of $\epsilon$ transitions after the entire input has been read. Formally, a run $p$ on $x \in A^\infty$ is a sequence of states such that there is a monotonically increasing sequence of indices $j_0 = 0 < j_1 < \ldots < j_i < \ldots$ of length exactly $|x|$ such that $p(0) \in Q_0$ and for all $0 \le k < |p|$ either $k + 1 = j_i + 1$ for some $0 \le i < |x|$ and $p(k+1) \in \mathbf{n}[p(k), x(i)]$, or $k+1 \ne j_i + 1$ for any $i$ and $p(k+1) \in \mathbf{n}[p(k), \epsilon]$. The requirement that the sequence of $j_i$'s be the same length as $x$ ensures that every element of $x$ is read. A run is accepting if it is finite and $p(|p|) \in Q_F$ or if it is infinite and $\mathrm{In}(p) \cap Q_B \ne \emptyset$. Note that a finite sequence can be accepted via an infinite run.

A nondeterministic mixed automaton with $\epsilon$ transitions can be converted to a mixed automaton with no $\epsilon$ transitions by using the $\epsilon$-closure construction of Hopcroft and Ullman [44] although special attention must be paid to the sets of accepting states in the resulting automaton. Given an automaton $\mathcal{M}$, we define a function $\epsilon^+ : Q \to 2^Q$, which maps each $q_i \in Q$ to the set of all states that can be reached from $q_i$ via *one or more* $\epsilon$ transitions. Another function $\epsilon^* : Q \to 2^Q$ is defined so that $\epsilon^*(q_i) = \epsilon^+(q_i) \cup \{q_i\}$. We define an automaton $M'$ with $Q' = \{\epsilon^*(q_i) \mid q_i \in Q\}$ and $Q'_0 = \{\epsilon^*(q_i) \mid q_i \in Q_0\}$.

We define $\mathbf{n}' : Q' \times A \to 2^{Q'}$ so that

$$\mathbf{n}'(\epsilon^*(q_i), a) = \{\epsilon^*(q_j) \mid \exists q_k \in \epsilon^*(q_i) : q_j \in \mathbf{n}(q_k, a)\}.$$

The finitely accepting states of the automaton must include not only the finitely accepting states of the original, but also the *infinitely* accepting states in the original that can be reached from themselves via one or more epsilon transitions:

$$Q'_F = \{\epsilon^*(q_i) \mid \exists q_j \in \epsilon^*(q_i) : q_j \in Q_F \lor (q_j \in Q_B \cap \epsilon^+(q_j)).$$

Note that this construction does not increase the number of states in the automaton. It may increase the number of transitions, but the total number of transitions is bounded by $|Q|^2 \cdot |A|$.

Given a mixed automaton accepting a language $X \subseteq A^\infty$, there is a procedure to construct the automaton for $\mathrm{del}(D)(X)$, where $D \subseteq X$: replace all of the transitions on $D$-symbols by transitions on $\epsilon$, then convert to an $\epsilon$-free mixed automaton.

In order to describe several related mixed regular sets with a single state diagram, each state is labeled with the names of the sets for which it is an accepting state. There is a second label to say whether the state is finitely or infinitely accepting. The second character in the state label is "$*$" if it is a final state in a finite automaton, "$\omega$" if it is a final state in a Büchi automaton, or "$\infty$" if it is both. Hence, if a state is labeled "$S*F\omega$", any finite input having a run ending in that state is in $S$, while any infinite input having a run in which the state appears infinitely often is in $F$.

When $x$ is a member of $A^*$ and $X$ is a subset of $A^\infty$, we define the function $\mathbf{suf}(x, X)$ to be the set $\{y \mid xy \in X\}$. The following properties of $\mathbf{suf}(x, X)$ are easy to prove:

6.2 $$\mathbf{suf}(x, X) \cup \mathbf{suf}(x, Y) = \mathbf{suf}(x, X \cup Y)$$

6.3 $$\mathbf{suf}(x, X) \cap \mathbf{suf}(x, Y) = \mathbf{suf}(x, X \cap Y)$$

6.4 $$\mathbf{del}(D)[\mathbf{suf}(x, X)] \subseteq \mathbf{suf}[\mathbf{del}(D)(x), \mathbf{del}(D)(X)]$$

6.5 $$A^\infty - \mathbf{suf}(x, X) = \mathbf{suf}(x, A^\infty - X)$$

6.6 $$X \subseteq Y \quad \text{implies} \quad \mathbf{suf}(x, X) \subseteq \mathbf{suf}(x, Y)$$

If $X \subseteq A^\infty$, two finite sequences $x$ and $y$ (in $A^*$) are *suffix-equivalent in $X$* if $\mathbf{suf}(x, X) = \mathbf{suf}(y, X)$. The following two technical lemmas are used in the final section of this chapter (an equivalence relation of finite index has a finite number of equivalence classes):

**Lemma 6.5.** *Suffix equivalence in a regular set $X$ is an equivalence relation of finite index.*

**proof.** Suffix equivalence is obviously an equivalence relation. To see that it is of finite index, consider first the case where $X \subseteq A^*$ or $X \subseteq A^\omega$. There is a nondeterministic finite (or Büchi) automaton $\mathcal{M}$ accepting $X$. We say that a finite sequence $x$ of length $i$ *reaches* a set of states $R$ if $R = \{p(i) \mid p \text{ is a run of } \mathcal{M} \text{ on } x\}$. Clearly, if $x$ and $y$ reach the same sets of states in $\mathcal{M}$, $x$ and $y$ are suffix-equivalent in $X$: for any sequence $z$, $xz$ is accepted if and only $yz$ is accepted (the suffixes of the runs on $xz$ and $yz$ are identical). But there are only a finite number of distinct state sets, so the index of the equivalence relation is finite.

When $X$ contains both finite and infinite sequences, $\mathbf{suf}(x, X) = \mathbf{suf}(x, X \cap A^*) \cup \mathbf{suf}(x, X \cap A^\omega)$. It is clear that the index of suffix equivalence in $X$ is no greater than the product of the indices of suffix equivalence in $X \cap A^*$ and $X \cap A^\omega$. $\square$

**Lemma 6.6.** *Suffix equivalence is right-invariant with respect to concatenation.*

**proof.** An equivalence relation, $\sim$, on finite sequences is right invariant if for every $a \in A$, $x \sim y$ implies $xa \sim ya$. This is obviously true from the definition of **suf**. $\square$

It should be clear from this proof that the deterministic automaton resulting from the usual subset construction has the property that two finite sequences reach the same state if and only if they are suffix-equivalent. So an effective representation of the suffix-equivalent sequences can be found in time exponential in the size of $Q$.

## 6.4. Infinite Games

As we shall see below, *infinite games* are a central concept in the study of complete trace structures. This section defines infinite games and presents basic results relating to them.

We define an *infinite game* to be a triple $(X, Y, Z)$, where $X$ and $Y$ are any sets and $Z$ is a set of sequences in $(X \times Y)^\omega$. There are two *players* of the game, which we call *white* and *black*. $X$ is the set of *white moves*, $Y$ is the set of *black moves*, and $Z$ is the set of *black wins*. Each *play* of the game consists of a white move followed by a black move, formally, a pair in $X \times Y$. A *history* of a game is a sequence in $(X \times Y)^\omega$ (so $Z$ is a set of histories).

A history is a win for black if it is in $Z$, and a win for white otherwise. A *strategy* for black is a mapping from $X^{(*)}$ to $Y$. The sets of moves $X$ and $Y$ and a strategy $f$ for black defines the histories that can result if black plays *according to $f$*. If for every sequence $x_0, x_1, \ldots$ in $X^{(\omega)}$ it is true that

$$(x_0, f(x_0)), (x_1, f(x_0, x_1)), \ldots, (x_i, f(x_0, \ldots, x_i)) \ldots \in Z,$$

then $f$ *is a winning strategy for black*. If there *exists* a winning strategy for black in a particular game $G = (X, Y, Z)$, we say that $G$ is *winnable by black* (or *black-winnable*).

There are two obvious but useful observations to be made about black winnability:

(G1) If $(X, Y, Z)$ is winnable by black and $Z \subseteq Z'$, then $(X, Y, Z')$ is winnable by black also.

(G2) $(X, Y, (X \times Y)^\omega)$ is winnable by black.

A *game projection*, $\phi$, on an infinite game $G = (X, Y, Z)$ is a pair of functions $(\phi_W, \phi_B)$, where $\phi_W$ is any surjection from $X$ to some set $X'$, and $\phi_B$ is a surjection from $Y$ to $Y'$. $\phi$ is extended to pairs $(x, y)$ in $X \times Y$ so that $\phi(x, y) = (\phi_W(x), \phi_B(y))$. It is naturally extended to finite sequences (members of $(X \times Y)^*$) and is further extended to infinite sequences by defining $\phi(z) = \lim(\phi[\mathbf{pref}(z)])$ for every $z \in (X \times Y)^\omega$ (so

$\phi[(x_0, y_0)(x_1, y_1), \ldots] = x_0 y_0 x_1 y_1 \ldots)$. $\phi$ is further extended to sets. Finally $\phi(\mathcal{G})$ is defined to be the game $\mathcal{G}' = (\phi_W(X), \phi_B(Y), \phi(\mathcal{Z}))$.

Given a projection $\phi$, the game $(\phi_W^{-1}(X), \phi_B^{-1}(Y), \phi^{-1}(\mathcal{Z}))$ is the *inverse projection*, written $\phi^{-1}(\mathcal{G})$. If $\mathcal{G}' = \phi^{-1}(\mathcal{G})$, it is the game having the largest set of black wins such that $\phi(\mathcal{G}') = \mathcal{G}$. Intuitively, if a projection is not injective, it may map two histories in the original game to the same history in the image game. In case both black-wins and white-wins map to the same history, the image history is classified as a black-win, so projection is "biased" toward black. Inverse projection may split each player's moves into interchangeable copies, which would seem not to affect the winnability of the game by either player. It is not very surprising, then, that black-winnable games are closed under projection and inverse projection. These results are used to prove that complete trace structures are closed under **rename, hide,** and **inverse deletion** in Chapter 7.

**Lemma 6.7.** *If $\mathcal{G}$ is winnable by black and $\phi$ is a projection of $\mathcal{G}$, then $\phi(\mathcal{G})$ is winnable by black.*

**proof.** Let $\mathcal{G}' = \phi(\mathcal{G})$. We construct a winning strategy for black in $\mathcal{G}'$. Since $\mathcal{G}$ is winnable by black, there is a winning strategy $f$. For each $x' \in X' = \phi_W(X)$, we choose a distinguished representative in $\phi_W^{-1}(x')$. For every sequence $x_0', x_1', \ldots, x_i' \in X'^{(*)}$, we set $f'(x_0', x_1', \ldots, x_i')$ to $\phi_B[f(x_0, x_1, \ldots, x_i)]$, where $x_i$ is the representative of $\phi_W^{-1}(x_i')$.

Let $x_0, x_1, \ldots$ be any sequence in $X^{(\omega)}$. Since $f$ is a winning strategy for black, the sequence $(x_0, f(x_0))(x_1, f(x_0, x_1)) \ldots$ is a member of $\mathcal{Z}$, so

$$(x_0', f'(x_0'))(x_1', f'(x_0', x_1')) \ldots = \phi[(x_0, f(x_0))(x_1, f(x_0, x_1)) \ldots]$$
$$\in \phi(\mathcal{Z}) = \mathcal{Z}'.$$

Hence, $f'$ is a winning strategy for black in $\mathcal{G}'$, so $\mathcal{G}'$ is winnable by black.  □

**Lemma 6.8.** *If $\phi$ a game projection, and $\mathcal{G}$ is winnable by black, then $\phi^{-1}(\mathcal{G})$ is winnable by black.*

**proof.** Let $\mathcal{G}' = \phi^{-1}(\mathcal{G})$. Since $\mathcal{G}$ is winnable by black, there is some winning strategy $f$. For each $y \in Y$ we choose a representative $y' \in \phi_B^{-1}(y)$. For any $x_0', x_1', \ldots, x_i' \in X'^{(*)}$, let $x_j = \phi_W(x_j')$ for all $0 \le j \le i$. We set $f'(x_0', x_1', \ldots, x_i') = y_i'$ where $y_i'$ is the representative of $\phi_B^{-1}(f[x_0, x_1, \ldots, x_i])$. Note that $\phi_B[f'(x_0', x_1', \ldots x_i')] = f(x_0, x_1, \ldots x_i)$. So

$$\phi[(x_0', f'(x_0'))(x_1', f'(x_0', x_1')) \ldots] = (x_0, f(x_0))(x_1, f(x_0, x_1)) \ldots$$
$$\in \mathcal{Z}.$$

Hence $(x_0', f'(x_0'))(x_1', f'(x_1')) \ldots \in \phi^{-1}(\mathcal{Z}) = \mathcal{Z}'$. Hence $f'$ is a winning strategy for black in $\mathcal{G}'$.  □

## 6.5. Decidability of Regular Games

A necessary but not sufficient condition for fully automatic verification is that all the problems involved be decidable. This section proves the decidability of the particular class of games of interest in Chapter 7.

We are interested in games in which the moves of each player are sequences. Formally, $X$ and $Y$ are subsets of $A^*$. For trace structures, we need to transform a set of *traces*, $Z \subseteq (XY)^\omega$ into a set of *histories*, $\mathcal{Z} \subseteq (X \times Y)^\omega$. $\mathbf{flat}(X, Y) : (X \times Y)^* \to (XY)^*$ is defined to map every finite sequence $(x_0, y_0)(x_1, y_1) \ldots (x_i, y_i)$ to the sequence $x_0 y_0 x_1 y_1 \ldots x_i y_i$. (The $(X, Y)$ arguments are usually omitted below, since $X$ and $Y$ are apparent from the context.) $\mathbf{flat}$ respects prefixes, so it can be extended to infinite sequences $\mathbf{z} \in (X \times Y)^\omega$ by $\mathbf{flat}(\mathbf{z}) = \lim(\mathbf{flat}[\mathbf{pref}(\mathbf{z})])$. We further naturally extend $\mathbf{flat}$ to sets (subsets of $(X \times Y)^\omega$). $\mathbf{flat}$ is used to convert a set of histories $\mathcal{Z}$ to a set of traces $Z$.

The inverse image of $\mathbf{flat}$ converts sets of traces to sets of histories: if $Z \subseteq (XY)^\omega$ is a set of traces, $\mathbf{flat}(X, Y)^{-1}(Z)$ is the set of all histories that can be flattened into members of $Z$.

For complete trace structures, we are interested in games $\mathcal{G}$ in which the move sets $X$ and $Y$ are regular sets and $\mathcal{Z} = \mathbf{flat}^{-1}(Z)$ for some $\omega$-regular set $Z$ (which must be a subset of $(XY)^\omega$). The problem of whether $\mathcal{G}$ is winnable by black is *decidable* for this class of games, if $X$, $Y$, and $\mathcal{Z}$ are represented in some effective form, such as mixed finite automata. The rest of this section is devoted to a proof of this claim. This is the most difficult result of the thesis. The reader who is prepared to believe it without proof should consider moving on to the next chapter.

**Theorem 6.1.** *It is decidable whether the infinite game $\mathcal{G} = (X, Y, \mathcal{Z})$ is winnable by black, if $X$ and $Y$ are regular sets and and $\mathcal{Z} = \mathbf{flat}^{-1}(Z)$ for some $\omega$-regular set $Z$.*

The approach is to reduce this problem to the Church's *solvability problem* [26], which has been shown to be decidable by Büchi and Landweber [21,75]. Rabin has also given a decision procedure for this problem, based on tree automata [65].

An instance of the *finite-state solvability problem* is a game $\mathcal{G} = (X, Y, \mathcal{Z})$, where $X$ and $Y$ are *finite sets* and $\mathcal{Z}$ is an $\omega$-*regular* subset of $(X \times Y)^\omega$. We assume that $\mathcal{Z}$ is given as a Muller automaton (there are procedures for converting $\omega$-regular expressions or Büchi automata to equivalent Muller automata [24,32]). The solvability problem is to determine whether $\mathcal{G}$ is winnable by black. $\mathcal{G}$ is not necessarily a solvability problem because $X$ and $Y$ can be infinite (and usually are in our application). We provide a procedure for reducing such game to an equivalent finite-state solvability problem.

The theorem is proved in the following lemmas. Throughout the rest of this section, let $A$ be a finite set which contains no sequences, let $X$ and $Y$ be regular subsets of $A^*$, let $Z$ be a mixed-regular subset of $(X \times Y)^\omega$, and let $Z' = \mathbf{flat}(X, Y)^{-1}(Z)$.

The central idea in the proof is to transform $X$ and $Y$ to finite sets $X'$ and $Y'$ by grouping their members into a finite set of equivalence classes.

Sistla, Vardi, and Wolper [71], proved that for every $\omega$-regular language, $W$, there is a finite partition $\Gamma$ of $A^+$ such that

$$W = \bigcup \{ PQ^\omega \mid P, Q \in \Gamma \wedge PQ^\omega \cap W \neq \emptyset \}.$$

This implies that for every $P, Q \in \Gamma$, either $PQ^\omega \cap W = \emptyset$ or $PQ^\omega \subseteq W$. Also, for every $P, Q \in \Gamma$, there exists $R \in \Gamma$ such that $PQ \subseteq R$. There is a procedure to compute $\Gamma$, given a Büchi automaton that accepts $W$. Every member of $\Gamma$ is a regular set. The procedure yields a family finite automata accepting the individual members of $\Gamma$.

We extend this result to mixed regular languages.

**Lemma 6.9.**  *There is a partition $\Gamma$ of $Z$ such that*

$$Z = \bigcup \{ PQ^\omega \mid P, Q \in \Gamma \wedge PQ^\omega \cap Z \neq \emptyset \}.$$

**proof.**  Let $Z_{\text{fin}} = Z \cap A^*$ and $Z_{\text{inf}} = Z \cap A^\omega$, and let $\Gamma'$ be the partition of $Z_{\text{inf}}$ as above. Let $\Gamma_1$ be $\{ P' \cap Z_{\text{fin}} \mid P' \in \Gamma' \}$, $\Gamma_2$ be $\{ P' \cap (A^+ - Z_{\text{fin}}) \mid P' \in \Gamma' \}$, and let $\Gamma = \Gamma_1 \cup \Gamma_2 \cup \{\{\epsilon\}\}$.

First, we show that for every $P, Q \in \Gamma$, either $PQ^\omega \subseteq Z$ or $PQ^\omega \cap Z = \emptyset$. This is clearly true for $P, Q \in \Gamma_1 \cup \Gamma_2$, since there are $P', Q' \in \Gamma'$ such that $P \subseteq P'$ and $Q \subseteq Q'$, and (by the properties of $\Gamma'$) either $P'Q'^\omega \subseteq Z$ or $P'Q'^\omega \cap Z = \emptyset$. There are three remaining cases. First, suppose $P = Q = \{\epsilon\}$. Then $PQ^\omega = \{\epsilon\}$, which is a subset of $Z$ if $\epsilon \in Z$ or disjoint from $Z$ otherwise. If $P = \{\epsilon\} \neq Q$, $PQ^\omega = Q^\omega = QQ^\omega$ which has already been considered in the first case. Finally, if $P \neq \{\epsilon\} = Q$, then $PQ^\omega = P$. If $P \in \Gamma_1$, $PQ^\omega \subseteq Z$ and if $P \in \Gamma_2$, $PQ^\omega \cap Z = \emptyset$.

Restating this, if $PQ^\omega \cap Z \neq \emptyset$, then $PQ^\omega \subseteq Z$. This implies that

$$\bigcup \{ PQ^\omega \mid P, Q \in \Gamma \wedge PQ^\omega \cap Z \neq \emptyset \} \subseteq Z.$$

To prove inclusion in the other direction, first, note $\Gamma_1 \cup \Gamma_2$ is a partition of $A^+$. It is obvious that $\bigcup(\Gamma_1 \cup \Gamma_2) = A^+$ from the definitions of $\Gamma_1$ and $\Gamma_2$ and the fact that $\bigcup \Gamma' = A^+$. Members of $\Gamma_1 \cup \Gamma_2$ are pairwise disjoint: consider any $P, Q$ in $\Gamma_1 \cup \Gamma_2$. If

$P$ and $Q$ are both in $\Gamma_1$ or both in $\Gamma_2$, they are disjoint because the members of $\Gamma'$ are disjoint; if $P \in \Gamma_1$ and $Q \in \Gamma_2$, they are disjoint because $Z_{\text{fin}}$ is disjoint from $A^+ - Z_{\text{fin}}$.

According to Lemma 2.3 of Sistla, Vardi, and Wolper's paper [71], if $\Gamma$ is a partition of $A^+$, then $\bigcup \{PQ^\omega \mid P, Q \in \Gamma\} = A^\omega$. This implies that

$$Z_{\text{inf}} \subseteq \bigcup \{PQ^\omega \mid (P, Q \in \Gamma_1 \cup \Gamma_2) \wedge (PQ^\omega \cap Z_{\text{inf}} \neq \emptyset)\}.$$

Also, it is clear that $Z_{\text{fin}} \subseteq \bigcup \{P(\{\epsilon\})^\omega \mid P \in \Gamma_1 \cup \{\{\epsilon\}\}\}$, so

$$Z \subseteq \bigcup \{PQ^\omega \mid P, Q \in \Gamma \wedge PQ^\omega \cap Z \neq \emptyset\}.$$

□

For the remainder of the section, let $\Gamma$ be a partition of $Z$, as in Lemma 6.9. We now prove that there is an instance of the solvability problem, $\mathcal{G}'$, such that $\mathcal{G}'$ is winnable by black if and only if $\mathcal{G}$ is. We define a game projection $\phi$: for any $x \in X$, we set $\phi_W(x) = P$, where $P$ is the unique member of $\Gamma$ such that $x \in P$. Similarly, for each $y \in Y$, we set $\phi_B(y) = Q$, where $Q \in \Gamma$ and $y \in Q$. The reduced game $\mathcal{G}'$ is $\phi(\mathcal{G})$ (so $X' = \phi(X)$, $Y' = \phi(Y)$, and $Z' = \phi(Z)$).

It is immediate from Lemma 6.7 that if $\mathcal{G}$ is winnable by black, so is $\mathcal{G}'$. To prove the converse, we show that $\mathcal{G} = \phi^{-1}(\mathcal{G}')$; this requires proving several lemmas.

**Lemma 6.10.** *If $P_0, P_1, \ldots$ and $Q_0, Q_1, \ldots$ are infinite sequences of subsets of $A^*$ and there is an infinite strictly increasing sequence of numbers $0 = i_0 < i_1 < \ldots$ such that $P_{i_j} P_{i_j+1} \ldots P_{i_{j+1}-1} \subseteq Q_j$ for all $j \in \omega$, then $P_0 P_1 \ldots \subseteq Q_0 Q_1 \ldots$.*

**proof.** Let $x$ be any sequence in $P_0 P_1 \ldots$. By definition, $x$ is the concatenation of some sequence $x_0, x_1 \ldots$ elementwise contained in $P_0, P_1 \ldots$. Let $y_0 y_1 \ldots$ be the sequence such that $y_j = x_{i_j} x_{i_j+1} \ldots x_{i_{j+1}-1}$ for all $j \in \omega$. Then for all $j \in \omega$, $y_j \in Q_j$, so $y_0 y_1 \ldots$ is in $Q_0 Q_1 \ldots$. But obviously, $y_0 y_1 \ldots = x$, so $P_0 P_1 \ldots \subseteq Q_0 Q_1 \ldots$.   □

**Lemma 6.11.** *For every infinite sequence $(P_0, Q_0)(P_1, Q_1) \ldots$ in $(X' \times Y')^\omega$, there is an infinite increasing sequence of indices $i_0 = 0 < i_1 < i_2 < \ldots$ such that there is a $P \in \Gamma$ such that $P_0 Q_0 \ldots P_{i_1-1} Q_{i_1-1} \subseteq P$ and there is a $Q \in \Gamma$ such that $P_{i_j} Q_{i_j} \ldots P_{i_{j+1}-1} Q_{i_{j+1}-1} \subseteq Q$ for all $j \geq 1$.*

**proof.** The proof of this theorem closely resembles the proof of Lemma 1 in a paper of Büchi's [20] and of Lemma 2.3 of Sistla, Vardi, and Wolper's paper (used in the proof of Lemma 6.9). The proof uses an infinite version of Ramsey's Theorem, which states that any finite partition of the unordered pairs of members of $\omega$ contains all the unordered pairs of some infinite subset of $\omega$.

Any $\omega$-sequence $\mathbf{z} = (P_0, Q_0)(P_1, Q_1)\ldots$ defines such a partition: every unordered pair of numbers $\{m, n\}$ $(m < n)$ uniquely delimits a subsequence $(P_m, Q_m), \ldots, (P_{n-1}, Q_{n-1})$. $\Gamma$ defines a finite partition of the unordered pairs, because for every pair $(m, n)$, $P_m Q_m \ldots P_{n-1} Q_{n-1} \subseteq R$ for exactly one $R \in \Gamma$. By Ramsey's Theorem, one member of this partition includes the set of all unordered pairs of some infinite subset of $\omega$. We call the elements of this set $i_1$, $i_2$, and so on, in ascending order $(i_1 < i_2 < \ldots)$. There exists some $Q \in \Gamma$ such that each subsequence has $P_{k_j} Q_{k_j} \ldots P_{i_{j+1}-1} Q_{i_{j+1}-1} \subseteq Q$.

As a special case, $P_0 Q_0 \ldots P_{i_1-1} Q_{i_1-1}$ is in some $P \in \Gamma$ (unless $i_1 = 0$, in which case we set $P = Q$, and rename $i_j$ to $i_{j-1}$ for all $j \in \omega$). Hence, there exist $P, Q \in \Gamma$ such that $P_0 Q_0 P_1 Q_1 \ldots \subseteq PQ^\omega$. $\square$

**Lemma 6.12.** *For every sequence* $(P_0, Q_0)(P_1, Q_1)\ldots \in \mathcal{Z}'$, $P_0 Q_0 P_1 Q_1 \ldots \subseteq Z$.

**proof.** By Lemma 6.11, there exist $P, Q \in \Gamma$ such that $P_0 Q_0 P_1 Q_1 \subseteq PQ^\omega$. By the definition of $\mathcal{Z}'$, there is some $(x_0, y_0)(x_1, y_1)\ldots \in \mathcal{Z}$ such that $x_i \in P_i$ and $y_i \in Q_i$ for all $i \in \omega$, so $x_0 y_0 x_1 y_1 \ldots \in PQ^\omega$. By the definition of $\mathcal{Z}$, $x_0 y_0 x_1 y_1 \ldots \in Z$, so $PQ^\omega \cap Z \neq \emptyset$. But by the properties of $\Gamma$ (Lemma 6.9), $PQ^\omega \subseteq Z$. Hence, $P_0 Q_0 P_1 Q_1 \ldots \subseteq PQ^\omega \subseteq Z$. $\square$

**Lemma 6.13.** $\mathcal{G} = \phi^{-1}(\mathcal{G}')$.

**proof.** The only non-obvious requirement is that $\mathcal{Z} = \phi^{-1}(\mathcal{Z}')$. By the properties of inverse images, $\mathcal{Z} \subseteq \phi^{-1}(\mathcal{Z}')$, so we need only prove inclusion in the opposite direction. Let $(x_0, y_0)(x_1, y_1)\ldots$ be any member of $\phi^{-1}(\mathcal{Z}')$. By the definition of $\phi$, there exists some $(P_0, Q_0)(P_1, Q_1)\ldots \in \mathcal{Z}'$ such that $x_i \in P_i$ and $y_i \in Q_i$ for all $i \in \omega$. By Lemma 6.12, $P_0 Q_0 P_1 Q_1 \ldots \subseteq Z$, so $x_0 y_0 x_1 y_1 \ldots \in Z$. Hence, $(x_0, y_0)(x_1, y_1) \in \text{flat}^{-1}(Z) = \mathcal{Z}$, so $\phi^{-1}(\mathcal{Z}') \subseteq \mathcal{Z}$. $\square$

By this lemma, $\mathcal{Z}$ is an inverse game projection of $\mathcal{Z}'$. By Lemma 6.8, if $\mathcal{Z}'$ is winnable by black, $\mathcal{Z}$ is too. Hence, we have proved that $\mathcal{G}'$ is equivalent to $\mathcal{G}$ with respect to black-winnability. To complete the proof of the theorem, we must show that $\mathcal{G}'$ is an instance of the solvability problem. Specifically, it must be proved that $\mathcal{Z}'$ is regular and that there is an effective method to find an automaton accepting it, given a Büchi automaton accepting $Z$.

We proved in Lemma 6.12 that the concatenation of every sequence of sets in $\mathcal{Z}'$ is a subset of $Z$. Here we prove that $Z$ is equal to the union of concatenations of these sets.

**Lemma 6.14.** *For every $(P_0, Q_0) \subseteq (X \times Y)^\omega$ such that $P_0 Q_0 \ldots \subseteq Z$, $(P_0, Q_0) \ldots \in \mathcal{Z}'$*

**proof.** If $P_0 Q_0 P_1 Q_1 \ldots \subseteq Z$, then there is some $x_0 y_0 x_1 y_1 \ldots \in Z$ such that $x_i \in P_i$ and $y_i \in Q_i$ for all $i \in \omega$. So, by definition, $(x_0, y_0)(x_1, y_1) \ldots \in \mathcal{Z}$ and so $(P_0, Q_0)(P_1, Q_1) \ldots \in \mathcal{Z}'$, also by definition.  □

Together, Lemmas 6.12 and 6.14 give a direct characterization of $\mathcal{Z}'$ in terms of $Z$: $\mathcal{Z}'$ is exactly the set of sequences $(P_0, Q_0)(P_1, Q_1) \ldots$ such that $P_i, Q_i \in \Gamma$, $P_i \cap X \neq \emptyset$, $Q_i \cap Y \neq \emptyset$, and $P_0 Q_0 P_1 Q_1 \ldots \cap Z \neq \emptyset$.

To show that $\mathcal{Z}'$ is $\omega$-regular, we define a *factorization* of $P \in \Gamma$ to be a finite sequence $P_0, P_1 \ldots P_i \in \Gamma^{(+)}$ such that $P_0 P_1 \ldots P_i \subseteq P$.

**Lemma 6.15.** *If $P$ is any member of $\Gamma$, the set of all factorizations of $P$ is regular.*

**proof.** Let the members of $\Gamma$ be $R_i$, and let the set of all factorizations of $R_i$ be $\mathcal{F}_i$. Clearly the sequence $R_i$ (of length 1) is in $\mathcal{F}_i$. If $R_j, R_k$ is a factorization of length 2, every sequence in $R_j \mathcal{F}_k$ is a factorization of $R_i$, also, because the concatenation is a subset of $R_j R_k \subseteq R_i$. Furthermore, if $R_j, R_{k_1}, R_{k_2}, \ldots, R_{k_l}$ is a factorization of $R_i$, then $R_{k_1} R_{k_2} \ldots R_{k_l}$ is a subset of some $R_k \in \Gamma$ (by the properties of $\Gamma$), so $R_{k_1}, R_{k_2}, \ldots, R_{k_l}$ is a member of some $\mathcal{F}_k$. Hence, every factorization of $R_i$ is either $R_i$ itself or a member of some $R_j \mathcal{F}_k$ where $R_j, R_k$ is a factorization of $R_i$.

Consequently the $\mathcal{F}_i$ for all $R_i \in \Gamma$ can be described as a right-linear grammar, where the $R_i$ are terminals and the $\mathcal{F}_i$ are non-terminals. The set of strings generated from $\mathcal{F}_i$ as the sentence symbol is the set of factorizations of $R_i$. The language of a right-linear grammar is always a regular set, so the lemma holds.  □

Lemma 6.15 gives an effective procedure for finding an automaton accepting each $\mathcal{F}_i$, since there is a procedure to find a finite automaton accepting the language of a right-linear grammar (Theorem 9.1 of Hopcroft and Ullman).

The actual sets of interest are more constrained than arbitrary factorizations: we want the $\mathcal{F}_i'$ of finite sequences of the form $(P_0, Q_0) \ldots (P_j, Q_j)$ in $(X' \times Y')^+$, where $P_0, Q_0, \ldots, P_j, Q_j$ is a factorization of $R_i$.

**Lemma 6.16.** *The $\mathcal{F}_i'$ are regular and effectively computable.*

**proof.** First, we find the set of factorizations $\mathcal{F}_i''$ of $R_i$ that are also of the form $P_0, Q_0, \ldots P_i, Q_i$, where $P_j \in X'$ and $Q_j \in Y'$ for $0 \leq j \leq i$. $\mathcal{F}_i''$ is the intersection of $\mathcal{F}_i$ with $(X'Y')^+$ (alternating sequences of members of $X'$ and $Y'$). A finite automaton for $\mathcal{F}_i''$ can be constructed by intersecting the automaton for $\mathcal{F}_i$ with an automaton accepting $(X'Y')^+$ (it has two states). Hence, $F_i''$ is regular and effectively computable by intersecting finite automata.

Let $h$ be the homomorphism that maps symbols $(P_i, Q_i)$ to the sequences $P_i, Q_i$. Then $\mathcal{F}_i' = h^{-1}(F_i')$. Since regular sets are closed under inverse homomorphism, $\mathcal{F}_i'$ is regular. Theorem 3.5 of Hopcroft and Ullman gives a construction for finding an automaton accepting the inverse homomorphic image of the language of any finite automaton. □

**Lemma 6.17.**  $\mathcal{Z}' = \{\mathcal{F}_k'\mathcal{F}_l'^{\omega} \mid R_k, R_l \in \Gamma \land R_k R_l^{\omega} \cap Z \neq \emptyset\}$

**proof.** Let $(P_0, Q_0), (P_1, Q_1), \ldots$ be any member of $\mathcal{F}_k'\mathcal{F}_l'^{\omega}$, where $R_k$ and $R_l$ are members of $\Gamma$ and $R_k R_l^{\omega} \cap Z \neq \emptyset$. Then $P_0 Q_0 P_1 Q_1 \ldots \subseteq R_k R_l^{\omega} \subseteq Z$ by the definition of the $\mathcal{F}_i'$ and the properties of $\Gamma$. But then by Lemma 6.14, $(P_0, Q_0)(P_1, Q_1) \ldots \subseteq \mathcal{Z}'$. Hence, $\{\mathcal{F}_k'\mathcal{F}_l'^{\omega} \mid R_k, R_l \in \Gamma \land R_k R_l^{\omega} \cap Z \neq \emptyset\} \subseteq \mathcal{Z}'$

Now let $(P_0, Q_0)(P_1, Q_1) \ldots$ be any member of $\mathcal{Z}'$. By Lemma 6.11, the sequence can be segmented into finite subsequences $(P_{i_j}, Q_{i_j}) \ldots (P_{i_{j+1}-1}, Q_{i_{j+1}-1})$ such that $P_0 Q_0 \ldots P_{i_1} Q_{i_1} \subseteq R_k$ for some $R_k \in \Gamma$, and $P_{i_j} Q_{i_j} \ldots P_{i_{j+1}-1} Q_{i_{j+1}-1} \subseteq R_l$ for some $R_l \in \Gamma$, for all $j \in \omega$. Hence, $(P_0, Q_0)(P_1, Q_1) \ldots \in \mathcal{F}_k'\mathcal{F}_l'^{\omega}$. Therefore, $\mathcal{Z}' \subseteq \{\mathcal{F}_k'\mathcal{F}_k'^{\omega} \mid R_k, R_l \in \Gamma \land R_k(R_l)^{\omega} \cap Z \neq \emptyset\}$. □

This completes the proof of Theorem 6.1.

Although the problem is decidable, the decision procedure falls short of practicality. First, the automaton for $\mathcal{Z}'$ may be have exponentially more states than the original automaton, because the alphabet of $\mathcal{Z}'$ can be exponentially larger than the number of states in the automaton accepting $Z$. Second, the decision procedure for the solvability problem operates on deterministic Muller automata, and the known procedures for transforming a nondeterministic Büchi automaton to an equivalent a deterministic Muller automaton [24,81] are at least doubly exponential in the worst case. Finally, Landweber and Büchi's decision procedure for the solvability problem adds at least another exponential to this.

There is potential room for improvement in the procedure. The solvability problem is PSPACE hard, since it can be reduced to the universality problem for nondeterministic Büchi automata, so it is unlikely that a subexponential algorithm can be found. However, this is the only lower bound on the complexity of the problem that is known to the author. If a singly exponential algorithm can be found for problem, the decision procedure would be comparable with other potentially exponential constructions that are necessary anyway, such as complementation of nondeterministic Büchi automata.

# Chapter 7

# Complete Trace Structures

## 7.1. Introduction

The theory of the previous chapters is very good at saying what circuits must not do (violate safety properties), but it has nothing to say about what they *must* do (liveness properties). This chapter presents an extended trace theory that captures general liveness properties as well as safety properties.

The most blatant example of the limited expressive and modeling power of prefix-closed trace structures is the universal do-nothing module of Section 3.4, which implements everything by the conservative approach of doing nothing at all. It should be possible to forbid this type of implementation. As another example, consider two types of non-inverting buffer. The first is the practical kind: if it is quiescent and it receives an input transition, it *eventually* produces an output transition, after an arbitrary time. The second is an unreliable buffer: when it receives an input, it may produce an output after a finite delay, or it may not. The sets of partial executions of these circuits are identical — prefix-closed trace structures cannot distinguish them.

The representations developed in this chapter are called *complete trace structures*.[*] Prefix-closed traces represent *partial* executions; to represent liveness properties, traces should represent *complete* executions. When a trace ends, no more transitions occur in the system. Sets of complete traces are not necessarily prefix-closed. To see how this enhances expressive power, suppose that when a circuit receives an $a$ input, it eventually produces a $b$ output (a liveness property). This can be represented by making the trace $ab$ possible, but making the trace $a$ impossible. In this case, the $P$ set is not prefix-closed.

---

[*] This term was borrowed from Kevin Van Horn [80], who called complete executions *complete traces*.

To say that the environment *must* send an *a* when the circuit outputs a *b*, a trace structure can have the trace *b* as a failure and *ba* as a success. In this case, $S$ is not prefix-closed.

One of the consequences of the complete execution interpretation is that *infinite* traces are needed to represent non-terminating executions. For example, one would like to say that if a buffer receives an infinite number of inputs, it produces an infinite number of outputs. Moreover, some circuits have *no* finite behaviors (for example, a ring oscillator). To model these infinite executions, trace sets are extended to include infinite traces. The properties of complete trace structures closely parallel the the properties of prefix-closed trace structures. However, some of the definitions and decision procedures become more difficult, particularly those relating to receptiveness and the canonical form for conformation equivalence.

The second section defines a semantics for asynchronous circuits based on complete trace structures. Various functions are extended to infinite sequences, the requirement of prefix-closure is eliminated, and a new definition of receptiveness is introduced which is based on infinite games. The third section shows that complete trace structures form a circuit algebra. The fourth section discusses conformation in the context of complete traces. Most of the section is devoted to the canonical form for conformation equivalence and simplifications to achieve it. The primary surprises in this section are that autofailure manifestation can be defined using infinite games, and that even in the canonical form the set of failures is not a function of the set of successes. Based on all of this, the mirror operation for complete trace structures is defined. The fifth section illustrates complete trace structures and conformation through examples. The sixth section considers complete trace structures as a lattice; another way in which the theory of complete trace structures diverges from prefix-closed trace structures is that the simple relationship between meet and join and union and intersection of the $P$ and $F$ sets does not hold. Finally, the last section discusses some of the problems that might arise in implementing the theory.

## 7.2.   Complete Trace Structures and Receptiveness

The definition of *complete trace structures* is quite similar to prefix-closed trace structures. A complete trace structure $T$ is a quadruple $(I, O, S, F)$, where $I$ and $O$ are finite disjoint subsets of *Wires*, $S$ and $F$ are mixed regular subsets of $A^\infty$ (where $A = I \cup O$). $P$ is defined to be $S \cup F$, and must be non-empty. $S$ and $P$ *need not* be prefix-closed; the intuitive meaning of these traces is that they are *complete*, not partial, executions of a circuit.

The final property of complete trace structures, *receptiveness*, is much more difficult for complete trace structures than for prefix-closed trace structures. As with prefix-closed trace structures, receptiveness means intuitively that a circuit cannot control its inputs. The set of possible traces must include all of the inputs that can be sent by another device, whether they are allowed or not. The definition is developed in the rest of this section.

Recall the definition of receptiveness for prefix-closed trace structures: $PI \subseteq P$. Without this requirement, a trace structure could constrain its input by having a trace $x \in P$ but not $xa \in P$ for some $a \in I$, which would prevent any circuit from sending it an $a$ after the trace $x$. A prefix-closed trace structure can fail to be receptive *only* by refusing to receive inputs.

Without the prefix-closure requirement, there are other ways for inputs to be controlled. As a simple example, if $x \notin P$ but $xa \in P$ for some $a \in I$, some circuit is *required* to send an $a$ after the trace $x$. For receptiveness in complete trace structures, it must both be possible to receive additional inputs and *not* receive more inputs. As another example, suppose that **pref**$[(ab)^*]$ is subset of $P$, but $(ab)^\omega$ is not. Then the trace structure can receive any *finite* number of inputs, but not an infinite number. From this example, it is clear that a definition of receptiveness cannot be phrased solely in terms of the finite traces in $P$ — the infinite traces must also be taken into account.

The circuit and its environment can be regarded as two entities that build up a trace by adding input and output symbols to it. This goes on forever (a finite trace is built if the circuit and environment only add $\epsilon$ to the trace after some finite time). There is another way to phrase receptiveness: no matter what input signals arrive there must be *something* that the output can do to cause the resulting trace to be in the possible set. If this is not true for some sequence of appended inputs, the trace structure will implicitly prevent those inputs, which violates receptiveness.

The previous paragraph describes an infinite two-player game: if the circuit tries to cause the trace to be in the $P$ set and the environment tries to cause it not to be in $P$, the circuit is receptive *only if there is a winning strategy for the circuit.* Several details need to be fleshed out to formalize this idea.

First, what should each player's moves be? Since no assumptions are made about the relative speeds of components, we should assume that environment can be arbitrarily fast. An arbitrarily fast environment is a worst-case adversary, since a fast environment can always act like a slow one, but a slow environment cannot act like a fast one. Hence, the environment should be allowed to add any number of inputs to the trace in a single turn. However, the environment cannot be *infinitely* faster than the circuit, so it should only be allowed to add a *finite* number of inputs in a move. This argument suggests that the set of input moves should be $I^*$. When the output gets a turn, it should be able to add at least one symbol to the trace. In fact, it should only be able to add one output since an arbitrarily fast environment might be able to slip an input between two successive outputs, no matter how close together the two outputs are. Of course, the output should be able to add $\epsilon$ to the trace, so the output moves should be $O \cup \{\epsilon\}$. The environment should get the first move; if it is arbitrarily fast, it can produce inputs before the circuit gets around to producing its first output. Hence, the environment plays white and the circuit plays black.

All that remains to complete the definition of an infinite game is to define the set of black wins. For receptiveness, the circuit (black) should win when the trace that is built is in $P$, so the set of black wins should be $\mathbf{flat}^{-1}(P)$. So the entire game is $\mathcal{G} = (I^*, O \cup \{\epsilon\}, \mathbf{flat}^{-1}(P))$, and the circuit is receptive if $\mathcal{G}$ is black-winnable.

This is not *quite* the definition of receptiveness, however. There is one remaining subtlety, which can be illustrated by an example. Consider a (peculiar) selector with finite traces $[(a(b + c))^*]$ and infinite traces $[(ab)^{\omega}]$. If this selector receives a finite number of inputs, it can produce $b$'s and $c$'s arbitrarily, but if it receives an infinite number if inputs, it must produce only $b$'s. This trace structure is either absurd or useless. In reality, the circuit has no way of knowing whether, after receiving the first $a$, it is going to get an infinite number of inputs. If it outputs a $c$, it constrains the input to be finite, in which case it is not receptive; if it only outputs $b$'s, even for finite inputs, there is no point in including $c$'s in the trace set. The finite traces $[(ab)^*]$ are the only ones that can really occur. The definition of receptiveness should require that the remainder of the $P$ set after *any* finite prefix of $P$ generate a black-winnable game.

It is now possible to give the full definition of receptiveness for complete trace structures: *a complete trace structure $T$ is receptive if and only if, for all $x \in \mathbf{pref}(P)$, $(I^*, O \cup \{\epsilon\}, \mathbf{flat}^{-1}[\mathbf{suf}(x, P)])$ is winnable by black.*

To decide the receptiveness of a trace structure, it is necessary to discover whether the game corresponding to every prefix is black-winnable. Fortunately, two prefixes generate the same games if they are suffix-equivalent, so by Lemma 6.5, there are only a finite number of games that must be checked for black-winnability.

This problem appears to have been identified in only one other place in the literature*. In a Masters thesis at Caltech, Van Horn proposed complete traces for a general semantics of concurrency [80]. Van Horn also encountered the problem of receptiveness, which he solved by, in essence, requiring that there be a program with appropriate input and output variables that can exhibit a subset of the specified trace set. Such a process bears

---

*Since this was written, Pnueli and Rosner [63] have written about necessary and sufficient conditions for a specification to be implementable, given pre-designated input and output variables. This problem is closely related to receptiveness; they also encountered the Church solvability problem.

some resemblance to a strategy. The question of decidability for finite-state systems was neither raised nor solved.

## 7.3. Complete Trace Structures are a Circuit Algebra

The definitions of the operations **compose, hide,** and **rename** are exactly as for prefix-closed trace structures (see Section 3.3). However, both deletion and renaming functions are extended to infinite sequences as in Chapter 6. To illustrate the effect of the extended composition operation, consider the trace structures $(\emptyset, \{a\}, a^\omega, \emptyset)$ and $(\emptyset, \{b\}, b^\omega, \emptyset)$. The composition of these structures is $(\emptyset, \{a, b\}, (a^+ b^+ + b^+ a^+)^\omega, \emptyset)$ (the traces in $(a+b)^\omega$ that have an infinite number of both $a$'s and $b$'s). This models the reality that both circuits run at a finite speed.

The remainder of this section is devoted to proving the following result:

**Theorem 7.1.** *Complete trace structures are a circuit algebra.*

The reader who is prepared to accept this claim without further discussion may want to skip to the next section.

First, we prove that complete trace structures are closed under the circuit algebra operations. Like regular sets, mixed regular sets are closed under deletion, inverse deletion, intersection, and bijective renaming, so the only question with respect to closure is whether the result of any operation is receptive.

**Lemma 7.1.** *Complete trace structures are closed under inverse deletion.*

**proof.** Let $T$ be any complete trace structure, let $T' = \mathbf{del}(D)^{-1}(T)$, let $x'$ be any prefix of $P'$, let $x = \mathbf{del}(D)(x')$, and let $Z = \mathbf{flat}^{-1}[\mathbf{suf}(x, P)]$. $T$ is receptive and $x \in \mathbf{pref}(P)$, so there is a winning strategy $f$ for black. Let $\phi$ be the game projection such that $\phi_W = \mathbf{del}(D)$ and $\phi_B$ is the identity function. Since $T$ is receptive, the game $(I^*, O \cup \{\epsilon\}, Z)$ is winnable by black, so by Lemma 6.8, the game

$$(I'^*, O \cup \{\epsilon\}, \phi^{-1}(Z)) = (\phi_W^{-1}(I^*), \phi_B^{-1}[O \cup \{\epsilon\}], \phi^{-1}(Z))$$

is winnable by black, also.

Obviously, $\phi^{-1}(Z) \subseteq \mathbf{flat}^{-1}(\mathbf{del}(D)^{-1}[\mathbf{suf}(x, P)])$. Furthermore, we claim that

$$\mathbf{del}(D)^{-1}[\mathbf{suf}(x, P)] \subseteq \mathbf{suf}(x', P').$$

If $y'$ is any sequence in $\mathbf{del}(D)^{-1}[\mathbf{suf}(x, P)]$, there is a $y = \mathbf{del}(D)(y')$ such that $xy \in P$. But $\mathbf{del}(D)(x'y') = xy \in P$ so $x'y' \in P' = \mathbf{del}(D)^{-1}(P)$, so, $y' \in \mathbf{suf}(x', P')$.

$\mathbf{flat}^{-1}$ preserves set inclusion, so $\phi^{-1}(Z) \subseteq \mathbf{flat}^{-1}[\mathbf{suf}(x', P')]$. By property G1, $(I'^*, O \cup \{\epsilon\}, \mathbf{flat}^{-1}[\mathbf{suf}(x', P')])$ is winnable by black. Hence, $T'$ is receptive. $\square$

**Lemma 7.2.** *Complete trace structures are closed under intersection.*

**proof.** Let $T$ and $T'$ be any complete traces such that $A = A'$ and $O \cap O' = \emptyset$, and let $T'' = T \cap T'$. Let $x$ be any member of $\mathbf{pref}(P'')$, and let the games

$$\mathcal{G} = (I^*, O \cup \{\epsilon\}, \mathbf{flat}^{-1}[\mathbf{suf}(x, P)])$$
$$\mathcal{G}' = (I'^*, O' \cup \{\epsilon\}, \mathbf{flat}^{-1}[\mathbf{suf}(x, P')]), \text{ and}$$
$$\mathcal{G}'' = (I''^*, O'' \cup \{\epsilon\}, \mathbf{flat}^{-1}[\mathbf{suf}(x, P'')]).$$

Since $T$ and $T'$ are receptive and $x \in \mathbf{pref}(P) \cap \mathbf{pref}(P')$ by property 3.8, there are winning strategies for black in $\mathcal{G}$ and $\mathcal{G}'$, which we call $f$ and $f'$.

The main idea is to combine $f$ and $f'$ by alternating back and forth between them. If the resulting history is flattened, it can be segmented in different ways so that it appears to be a black win in either game. We define a strategy $f''$ for any finite sequence $x''_0, x''_1, \ldots x''_j$. The definition is by induction on the length of the sequence; it constructs two auxiliary sequences:

$$x_0 \triangleq x''_0,$$
$$f''(x''_0, \ldots, x''_{2i}) \triangleq f(x_0, \ldots, x_i),$$
$$x'_i \triangleq x''_{2i} f''(x''_0, x''_1, \ldots, x''_{2i}) x''_{2i+1},$$
$$f''(x''_0, \ldots, x''_{2i+1}) \triangleq f'(x'_0, \ldots, x'_i), \text{ and}$$
$$x_{i+1} \triangleq x''_{2i+1} f''(x''_0, x''_1, \ldots, x''_{2i+1}) x''_{2i+2}.$$

To see that $f''$ is a winning strategy for black in $\mathcal{G}''$, let $x''_0, x''_1, \ldots$ be any infinite sequence in $X''^{(\omega)}$. Then

$$\mathbf{flat}[(x''_0, f''(x''_0))(x''_1, f''(x''_0, x''_1)) \ldots] = x''_0 f''(x''_0) x''_1 f''(x''_0, x''_1) \ldots$$
$$= x_0 f(x_0) x_1 f(x_0, x_1) \ldots,$$

where $x_0 = x''_0$ and $x_{i+1} = x''_{2i+1} f''(x''_0, x''_1, \ldots, x''_{2i+1}) x''_{2i+2}$. Since $f$ is a winning strategy for black in $\mathcal{G}$, $(x_0, f(x_0))(x_1, f(x_0, x_1)) \ldots \in \mathcal{Z} = \mathbf{flat}^{-1}[\mathbf{suf}(x, P)]$, so $x_0 f(x_0) x_1 f(x_0, x_1) \ldots \in \mathbf{suf}(x, P)$. Similarly,

$$x''_0 f''(x''_0) x''_1 f''(x''_0, x''_1) \ldots = x'_0 f'(x'_0) x'_1 f'(x'_0, x'_1) \ldots$$
$$\in \mathbf{suf}(x, P'),$$

so $x_0'' f''(x_0'') x_1'' f''(x_0'', x_1'') \ldots \in \mathbf{suf}(x, P) \cap \mathbf{suf}(x, P')$. By property 6.3,

$$\mathbf{suf}(x, P) \cap \mathbf{suf}(x, P') = \mathbf{suf}(x, P''),$$

so $\mathbf{flat}[(x_0'', f(x_0''))(x_1'', f(x_0'', x_1'')) \ldots] \in \mathbf{suf}(x, P'')$. Hence,

$$(x_0'', f(x_0''))(x_1'', f(x_0'', x_1'')) \ldots \in \mathbf{flat}^{-1}[\mathbf{suf}(x, P'')],$$

which is the set of black wins for $\mathcal{G}''$. Thus, $\mathcal{G}''$ is winnable by black, so $T''$ is receptive.
□

**Corollary.**  *Complete trace structures are closed under composition.*

**Lemma 7.3.**  *Complete trace structures are closed under hiding.*

**proof.**  Let $T$ be any complete trace structure, let $T'$ be $\mathrm{hide}(D)(T)$ where $D$ is any subset of $O$, and let $x'$ be any member of $P'$. For every $x' \in \mathbf{del}(D)(P)$, there is an $x \in P$ such that $\mathbf{del}(D)(x) = x'$. $T$ is receptive, so the game $\mathcal{G} = (I^*, O \cup \{\epsilon\}, \mathcal{Z})$, where $\mathcal{Z} = \mathbf{flat}^{-1}[\mathbf{suf}(x, P)]$, is winnable by black. Let $\phi$ be the game projection such that $\phi_W$ is the identity function and $\phi_B = \mathbf{del}(D)$. By Lemma 6.7, the game $\phi(\mathcal{G}) = (I^*, O - D \cup \{\epsilon\}, \phi(\mathcal{Z}))$ is winnable by black. It is obvious that $\phi(\mathcal{Z}) \subseteq \mathbf{flat}^{-1}(\mathbf{del}(D)[\mathbf{suf}(x, P)])$; by property 6.4, $\mathbf{del}(D)[\mathbf{suf}(x, P)] \subseteq \mathbf{suf}(x', P')$, so $\phi(\mathcal{Z}) \subseteq \mathbf{flat}^{-1}[\mathbf{suf}(x', P')]$. Hence, by property G1, the game $(I^*, O - D \cup \{\epsilon\}, \mathbf{flat}^{-1}[\mathbf{suf}(x', P')])$ is winnable by black, so $T$ is receptive.  □

**Lemma 7.4.**  *Complete trace structures are closed under renaming.*

**proof.**  Let $T$ be any trace structure and let $\mathbf{r}$ be an appropriate renaming function. We define $\mathbf{r}_W = \mathbf{r}|_I$ and $\mathbf{r}_B = \mathbf{r}|_O$. If these functions are extended to sequences, we can define a game projection $\phi$ such that $\phi_W = \mathbf{r}_W$ and $\phi_B = \mathbf{r}_B$. The rest of the proof is obvious.  □

The proofs of circuit algebra laws C1 through C9 are all slight modifications of the proofs of the corresponding properties in Chapter 3. In a number of places, $A^\infty$ should be substituted for $A^*$. The only other changes occur when induction on sequences is used: once a property has been proved by induction on finite sequences, it must then be proved for infinite sequences. In general, these proofs are not sufficiently interesting to repeat. As a representative example, we extend Lemma 3.5 so that it applies to complete traces:

**Lemma 7.5.** *For any sequence $x$ in $(A \cup D)^\infty$ and renaming function* **r** *with domain* $A \cup D$, $\mathbf{r}[\mathbf{del}(D)(x)] = \mathbf{del}[\mathbf{r}(D)][\mathbf{r}(x)]$.

**proof.** Let $x$ be any member of $(A \cup D)^\infty$. By property 6.1, $x = \lim[\mathbf{pref}(x)]$, and by two applications of Lemma 6.4, $\mathbf{r}[\mathbf{del}(D)(\lim[\mathbf{pref}(x)])] = \lim[\mathbf{r}(\mathbf{del}(D)[\mathbf{pref}(x)])]$. By 3.5, this is equal to $\lim[\mathbf{del}[\mathbf{r}(D)](\mathbf{r}[\mathbf{pref}(x)])]$ (since $\mathbf{pref}(x)$ consists of finite traces only). By Lemma 6.4, this is equal to $\mathbf{del}[\mathbf{r}(D)][\mathbf{r}(\lim[\mathbf{pref}(x)])]$. Finally, by property 6.1, this is $\mathbf{del}[\mathbf{r}(D)][\mathbf{r}(x)]$.  □

## 7.4. Conformation

The definitions of *failure-free* and *conformation* $(\preceq)$ for complete trace structures are exactly as in Chapter 4. As with prefix-closed trace structures, complete trace structures can be reduced to a canonical form for conformation equivalence. However, some of the results are more complicated than the corresponding results from Chapter 4, and some do not hold at all. For example, the definition of autofailure is much more difficult (it uses infinite games). Furthermore, even when a complete trace structure has been reduced to canonical form, the $F$ set is *not* determined by the $S$ set. Finally, the mirror of a complete trace structure is more difficult both to define and to compute.

Intuitively, if the environment of the circuit cannot *guarantee* failure-free execution of a circuit after a particular partial trace $x$, $x$ is an autofailure (even if $x$ itself is a success). With prefix-closed trace structures, the environment could intervene to prevent a possible failure only if the circuit stopped and waited for inputs before proceeding to the failure. The environment could avert the failure by *not sending* an input that eventually leads to failure. Thus, an autofailure was any failure that could be reached by appending only outputs to $x$.

As one would expect, the increased expressiveness of complete trace structures introduces new kinds of failures. For example, if a trace $x$ is a failure but also a prefix to a success, the environment might be able to avert failure by *sending* inputs to the circuit. This situation does not arise in prefix-closed trace structures because $S$ is prefix-closed.

Autofailures can be defined by considering a game in which the circuit tries to fail and the environment tries to force the circuit to succeed. In this game, the environment cannot control the speed at which the circuit runs, so the circuit moves can be any finite sequence of output symbols (members of $O^*$), and the environment moves can be members of $I \cup \{\epsilon\}$. The environment loses this game if the circuit fails (the concatenation of the infinite history is a member of $F$). Equivalently, the environment wins if the history is a success (in $S$) or if the circuit "cheats" by playing a history that is not possible according to its specification (in $A^\infty - P$). Formally, a finite trace $x$ in **pref**$(P)$ is an autofailure if the infinite game $(O^*, I \cup \{\epsilon\}, \textbf{flat}^{-1}[A^\infty - \textbf{suf}(x, F)])$ is *not* black-winnable.

We define **af**$(T)$ to be the set of autofailures of $T$, according to the definition above. Call the result of applying autofailure manifestation $T'$; then $I' = I$, $O' = O$, $S' = S$, and $F' = F \cup \textbf{af}(T) \cdot A^\infty$. This transformation is effective: as with receptiveness, the prefixes of $P$ can be grouped into a finite number of classes by suffix equivalence, using the subset construction. Each one of these classes can be checked to see if it is an autofailure using the decision procedure of Chapter 6.

Autofailure manifestation can convert infinite failures into finite failures. For example, consider a complete trace structure $T$ which has inputs $a$ and $b$, outputs $c$ and $d$, and successful traces

$$S = ([(a + b)(c + d)]^* ([(a + b)c]^+ [(a + b)d]^+)^\omega$$

(inputs and outputs alternate, there must be an infinite number of inputs, and the circuit must produce an infinite number of both $c$'s and $d$'s). Suppose the failures include all possible sequences with only a finite number of inputs (so it is a failure for the environment to stop sending inputs), and also include the infinite traces in $[(a + b)(c + d)]^* bd[(a + b)d]^\omega$ (the outputs of the circuit must be fair). The finite prefixes in $[(a + b)(c + d)]^* b$ are autofailures.

Intuitively, if the environment ever sends a $b$ to the trace structure, the trace structure could send only $d$'s infinitely often, and fail as a consequence. Autofailure manifestation will make these finite prefixes into failure traces. In essence, part of the environmental assumption will be "the environment shall not send a $b$", which is a safety condition.

The second transformation, failure exclusion, is the same as in Chapter 4: the successes of the transformed structure are set to $S - F$; none of the other parts of the trace structure are changed. The simplifications produce complete trace structures that are conformation-equivalent to the originals.

We define a *canonical* complete trace structure $T = (I, O, S, F)$ to be a complete trace structure with the additional properties (1) $S \cap F = \emptyset$, and (2) $\textbf{af}(T) \cdot A^\infty \subseteq F$. As with prefix-closed trace structures, two canonical trace structures are equal if and only if they are conformation equivalent.

Unfortunately, unlike prefix-closed trace structures, the $S$ set does not completely determine the $F$ and $P$ sets in a canonical complete trace structure. As a simple example of this, consider a trace structure in which $a \in I$, $b \in O$, $xa, xb \in \textbf{pref}(S)$, but $x \notin S$. In this case, we may have either $x \in P$ (hence $x \in F$) or $x \notin P$. If $x \in F$, there is a requirement on the environment: it must eventually send an $a$; if $x \notin P$, it is a requirement on the circuit: if an $a$ does not arrive, the circuit eventually sends a $b$. Note, however, that since $S \cap F = \emptyset$ in a canonical complete trace structure, the sets $S$ and $P$ completely determine $F$ (which is $P - S$).

The mirror of a canonical trace structure is almost exactly the same as for prefix-closed trace structures: $T^M$ is defined to be $(O, I, S, A^\infty - F)$. As with prefix-closed trace structures, the mirror of a complete trace structure is the maximum circuit (under the conformation ordering) that can be composed with the original to yield a failure-free result.

This gives a procedure for testing whether $T \preceq T'$: first reduce $T'$ to canonical form, then check whether $T \parallel T'^M$ is failure-free. There is an algorithm to do this based on procedures to complement Büchi automata and test them for emptiness.

The rest of this section proves these claims. All of the results about conformation from Section 4.2, namely Lemmas 4.1 through 4.5, apply to complete trace structures as well as prefix-closed trace structures (the proofs do not need to be changed).

There is a counterpart to theorem 4.2 that applies to complete trace structures:

**Theorem 7.2.** *If $T$ is any complete trace structure, performing autofailure manifestation followed by failure exclusion yields a canonical complete trace structure that is conformation equivalent to $T$.*

The rest of this section proves this theorem.

**Lemma 7.6.** *Autofailure manifestation preserves complete trace structures.*

**proof.** Let $T'$ be the result of applying autofailure manifestation to a complete trace structure $T$. It is obvious that $T'$ is a complete trace structure, except perhaps for regularity of $S$ and $F$ and receptiveness. First, we prove that the set of autofailures of any complete trace structure is regular. If $x$ and $y$ are in $P$ and $\mathbf{suf}(x, F) = \mathbf{suf}(y, F)$ they are either both autofailures or both not (because they have the same games). Hence, by Lemmas 6.5 and 6.6, the set of autofailures is a union of classes of a right-invariant equivalence relation, and so by the Myhill-Nerode theorem, it is regular. $F'$ is the union of the regular sets $F$ and $\mathbf{af}(T)$, so it is regular. $S' = S$ is, of course, regular by hypothesis.

To see receptiveness, let $x$ be any prefix of $P'$. If $x \in \mathbf{pref}(P)$, the game

$$(I^*, O \cup \{\epsilon\}, \mathbf{flat}^{-1}[\mathbf{suf}(x, P)])$$

is winnable by black (by the assumed receptiveness of $T$). We know $\mathbf{suf}(x, P) \subseteq \mathbf{suf}(x, P')$ (by property 6.6), so by property G1, the game $(I^*, O \cup \{\epsilon\}, \mathbf{flat}^{-1}[\mathbf{suf}(x, P')])$ is black-winnable.

Otherwise, $x$ is in $\mathbf{pref}(P') - \mathbf{pref}(P)$. But then it is easy to see that $\mathbf{suf}(x, P') = A^\infty$. $(I^*, O \cup \{\epsilon\}, \mathbf{flat}^{-1}(A^\infty))$ is black-winnable by property G2. Hence, $T'$ is receptive.  $\square$

Lemma 4.6 from Section 4.3, which says that $P \subseteq P'$ and $F \subseteq F'$ implies that $T \preceq T'$, applies equally as well to complete trace structures.

**Lemma 7.7.**   *Autofailure manifestation preserves conformation equivalence.*

**proof.**   Let $T$ be any complete trace structure and let $T'$ be the result of applying autofailure manifestation to $T$. $P \subseteq P'$ and $F \subseteq F'$, so by Lemma 4.6, $T \preceq T'$. To prove $T' \preceq T$, let $T''$ be any complete trace structure such that $I'' = O$ and $O'' = I$, and suppose that $T \cap T''$ is failure-free.

We claim that no prefix of $P''$ can be an autofailure in $T$: let $x$ be any prefix of $P''$. Since $T \cap T''$ is failure-free, the set $F \cap P''$ is empty, so $P'' \subseteq A^\infty - F$. Furthermore, by property 6.5, $\mathbf{suf}(x, A^\infty - F) = A^\infty - \mathbf{suf}(x, F)$, so, since $\mathbf{suf}$ and $\mathbf{flat}^{-1}$ respect set inclusion, $\mathbf{flat}^{-1}[\mathbf{suf}(x, P'')] \subseteq \mathbf{flat}^{-1}[A^\infty - \mathbf{suf}(x, F)]$. By the receptiveness of $T''$, the game

$$(I''^*, O'' \cup \{\epsilon\}, \mathbf{flat}^{-1}[\mathbf{suf}(x, P'')]) = (O^*, I \cup \{\epsilon\}, \mathbf{flat}^{-1}[\mathbf{suf}(x, P'')])$$

is winnable by black. Hence, by property G1, $(O^*, I \cup \{\epsilon\}, \mathbf{flat}^{-1}[A^\infty - \mathbf{suf}(x, F)])$ is winnable by black for all $x \in \mathbf{pref}(P'')$. But then, by definition, $x$ cannot be a member of $\mathbf{af}(T)$. Hence, $\mathbf{pref}(P'') \cap \mathbf{af}(T) = \emptyset$, or, equivalently, $P'' \cap \mathbf{af}(T) \cdot A^\infty = \emptyset$.

By definition, $F' = F \cup \mathbf{af}(T) \cdot A^\infty$, so $F' \cap P'' = F \cap P'' = \emptyset$ (as noted above). Also, it is clear that $P' = P \cup \mathbf{af}(T) \cdot A^\infty$, so $P' \cap F'' = (P \cap F'') \cup [\mathbf{af}(T) \cdot A^\infty \cap F'']$. Since $T \cap T''$ was assumed to be failure-free, $P \cap F'' = \emptyset$, and since $F'' \subseteq P''$, $\mathbf{af}(T) \cdot A^\infty \cap F'' = \emptyset$, so $P' \cap F'' = \emptyset$, also.

Hence, the set of failures of $T' \cap T''$, which is $(F' \cap P'') \cup (P' \cap F'')$, is empty. So $T' \preceq T$.   $\square$

**Lemma 7.8.**   *Failure exclusion preserves complete trace structures.*

**proof.**   Obvious.   $\square$

**Lemma 7.9.**   *Failure exclusion preserves conformation equivalence.*

**proof.**   The $P$ and $F$ sets do not change, so this can be proved by two applications of Lemma 4.6.   $\square$

**Lemma 7.10.**   *Simplification reduces any complete trace structure to canonical form.*

**proof.**   Obvious.   $\square$

The next lemma parallels Lemma 4.18 of Section 4.4.

**Lemma 7.11.** *If $T$ is any canonical complete trace structure and $S \neq \emptyset$, then $T^M$ is also a canonical complete trace structure.*

**proof.** Let $T$ be any complete trace structure. The only real question about $T^M$ is whether it is receptive. Let $x$ be any prefix of $P^M$. If $x \in \text{pref}(P)$, we know it cannot be an autofailure in $T$, since in such a case $xA^\infty \subseteq F$, so $xA^\infty \cap (A^\infty - F) = xA^\infty \cap P^M = \emptyset$. Hence, by the definition of autofailure, if $x \in \text{pref}(P)$, $(O^*, I \cup \{\epsilon\}, A^\infty - \text{suf}(x, F)) = (I'^*, O' \cup \{\epsilon\}, \text{flat}^{-1}[\text{suf}(x, P^M)])$ is winnable by black.

Now suppose $x \notin \text{pref}(P)$, so $xA^\infty \subseteq A^\infty - P = F^M \subseteq P^M$. But then $\text{suf}(x, F) = A^\infty$; $(I^*, O \cup \{\epsilon\}, \text{flat}^{-1}(A^\infty))$ is certainly winnable by black, by property G2. Hence, $T^M$ is receptive. $\square$

The proofs of Lemmas 4.19 through 4.22 carry over to complete trace structures if $A^\infty$ is substituted for occurrences of $A^*$ (for example, $F^M = A^\infty - P$ instead of $F^M = A^* - P$). These lemmas constitute the proof of Theorem 4.3, which gave the mirror method for determining conformance.

Finally, Lemma 4.23, which proves $P \subseteq P'$ and $F \subseteq F'$ is a *necessary* condition for $T \preceq T'$, when $T$ and $T'$ are canonical complete trace structures, carries over from Section 4.4. It is used to prove Lemma 4.24, which asserts that if $T$ and $T'$ are canonical. $T \sim_C T'$ implies $T = T'$. The proof of Lemma 4.24 carries over, also. As with prefix-closed trace structures, operations can be defined which restore canonical forms using the simplifications. With such operators, canonical complete trace structures are a circuit algebra by the same argument as in Section 4.4, since the proof of Lemma 4.25 carries over without modification.

## 7.5.  Examples of Complete Trace Structures

### 7.5.1. Boolean Gates

When a NOR gate has both inputs 0 and output 0, the output will *eventually* go to 1. This is a liveness property that prefix-closed trace structures cannot model. Recall from Section 3.4 that a gate is stable when the value of the associated Boolean function is equal to its current output; otherwise, the gate is unstable. The additional property is that if a gate is continuously unstable, an output transition will *eventually* occur. It is easy to extend the general definition of trace structures for gates in Section 3.4 to include liveness.

A mixed automaton accepting the set of successes for a gate with Boolean function **b** has exactly the same set of states (**B**) and transitions as the automaton in Section 3.4. However, the set of finite accepting states $Q_F$ and the set of infinite accepting states $Q_B$

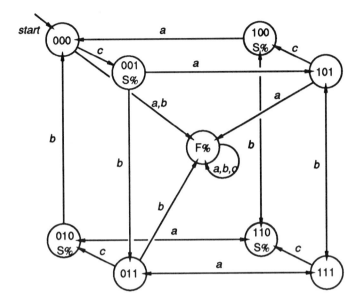

**Figure 7.1.**  Live NOR Gate

are the *stable* states (not every state in **B**, as in Section 3.4). In the automaton for the success set, the failure state $q_{fail}$ is in neither $Q_F$ nor $Q_B$.

A mixed automaton accepting the failures of the gate can be constructed by using the same states and transitions, but choosing $Q_F = Q_B = \{q_{fail}\}$. The requirements of this trace structure on the environment are essentially safety conditions (all of the failures are chokes).

Using this definition, the trace structure for any gate is receptive. For receptiveness, the circuit must have a strategy that keeps every infinite trace in $P$, no matter what inputs the environment sends. From any state **b**, a winning strategy for the circuit is to change the output whenever **b** is unstable, and to do nothing (play $\epsilon$) when **b** is stable. If the current state is $q_{fail}$, any output move will suffice. This is a winning strategy, because each time black gets a turn, it can force the automaton into a finitely and infinitely accepting state (for the $P$ set). If the trace is finite, the strategy makes sure that it ends up in a stable state, which is finitely accepting.

These trace structures are also canonical. In this case, the environment must have a strategy that keeps every trace in $A^{\infty} - F$, no matter what the circuit does. A winning strategy for the environment is to make $\epsilon$ moves whenever the moves reach an unstable state (essentially, waiting for the output to respond before sending another input). No history of a game played according to this strategy can reach $q_{\text{fail}}$.

Figure 7.1 shows an automaton that accepts the success or failure sets of a NOR gate. Note that this has exactly the same states and transitions as the corresponding automaton for the NOR in Section 3.4.

### 7.5.2. Ring Oscillator

Figure 7.2 shows an automaton for a live buffer or inverter. The initial state of the automaton depends on the initial wire values. For a buffer or inverter starting in a stable state, the set of successful traces is $(ab)^* + (ab)^\omega$. The failure traces are $(ab)^* aaA^\infty$ (the gate receives two inputs without an intervening output). Traces of the form $(ab)^* a$ are *impossible*, even though they are prefixes of possible traces. This models the inevitability of a $b$ transition once an $a$ has been received.

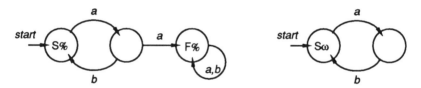

**Figure 7.2.**   Live Buffer or Inverter and Ring Oscillator

The trace structure for a ring oscillator can be constructed by composing this structure with an inverter with input $b$ and output $a$. The success set for such an inverter is $(ab)^* a + (ab)^\omega$, and the failure set is $(ab)^* abbA^\infty$. Since the alphabets match, the success set of the composition of these two trace structures is the intersection of the two success sets, which is $(ab)^\omega$. The two components have no *finite* traces in common, because all the buffer traces end with $b$ and all the inverter traces end with $a$. The only remaining trace is $(ab)^\omega$, which represents infinite oscillation. This correctly models the non-terminating behavior of a ring oscillator. Intuitively, none of the finite traces is possible because one of the components always insists on adding another output.

As in the example with prefix-closed trace structures, the composition has no failures. It is easy to see that the failures in each component are not possible in the other component.

### 7.5.3. Mutual Exclusion Element

There are at least two different interesting descriptions of mutual exclusion elements using complete traces, both of which have the same prefixes as the example in Section 3.4. First, the element should at least be live — it should eventually grant the resource to some client, and the release of the resource should eventually be acknowledged.

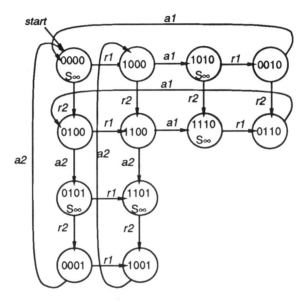

**Figure 7.3.**  Live Mutual Exclusion Element

Starting with the state graph for the mutual exclusion element of Section 3.4, we can require liveness by making it *impossible* for the automaton to wait forever in a state in which there are one or more outstanding requests, but none have been acknowledged. Similarly, it can be made impossible to wait forever in a state in which a release has been received but not acknowledged. Call these states *unstable*. If all but the unstable states are made finitely and infinitely accepting, the trace structure will have the desired properties. A state diagram is shown in Figure 7.3. The failures are not shown, to reduce clutter. The reader should imagine a single additional state marked "$F\infty$" with a self-loop transition on all symbols, as in the previous examples. Whenever there is no transition out of one of the states in the diagram for an input symbol, there must be a

transition on that symbol to the failure state. The arguments for why this structure is
receptive and canonical are similar to those for Boolean gates.

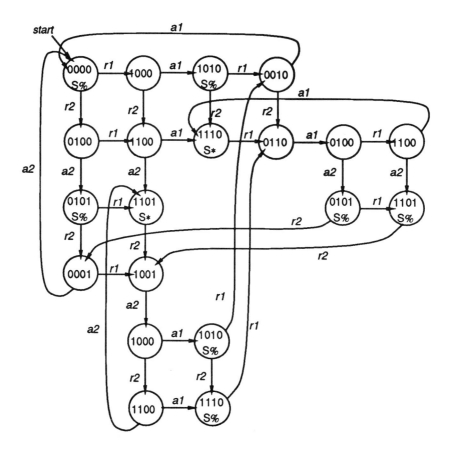

**Figure 7.4.**  Fair Mutual Exclusion Element

The first description of the mutual exclusion element allows *starvation*: If there are
infinitely many requests on one input, it is possible that requests on the other input
will never be granted. It is also possible by using complete trace structures to specify
unbounded *fairness*, so that a request pending on one of the inputs is eventually granted.
An automaton accepting appropriate trace sets is shown in Figure 7.4. It is similar to

the automaton of Figure 7.3, except it has been "unwound" somewhat, and the sets of accepting states are somewhat different. The failure state has been omitted, as in Figure 7.3, and it can be added in the same way. As with all the previous examples, there are no liveness requirements on the environment — any state that has no output transitions is finitely accepting for successes (it is permissible for the input to stop at that state).

To see why this mutual exclusion element is fair, consider a trace of the form

$$r2 \; (r1 \; a1 \; r1 \; a1)^{\omega},$$

which represents unfair behavior because the $r2$ request is ignored forever. Tracing the run through the automaton, we see that none of the four states that are repeated infinitely often is marked with $S\omega$ or $S\infty$. The only way such a state can be reached is for the symbol $a2$ to appear in the trace at some point. In general, whenever an $r2$ request occurs, the only way that an infinitely accepting state can be reached is for an $a2$ to occur. The situation with $r1$ and $a1$ is similar.

This trace structure is also both receptive and canonical. A receptiveness strategy would be for the output always to grant to the the earliest request that has not already been granted. A failure-avoidance strategy would be not to send inputs that lead directly to the failure state, as before.

### 7.5.4. Fork/Join

This example illustrates the differences between prefixed-closed and complete trace structures with respect to concurrency. It also gives a non-trivial example of an liveness requirement on a circuit environment.

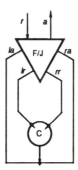

**Figure 7.5.** A Circuit that Distinguishes Fork/Join and Sequence

In Section 4.5, the discussion of the fork/join module concluded that it is impossible to specify, using prefix-closed trace structures, that an implementation must be *concurrent* — in particular, the sequence module conforms to the fork/join. Consider, however, the circuit of Figure 7.5, in which *rr* and *lr* are inputs to a C element, the output of which is connected to both *ra* and *la*. Using a fork/join, the circuit is live — if a transition *r* occurs, a transition on *a* eventually follows. However, if a sequence module is used instead of the fork/join, the circuit deadlocks after sending *lr*.

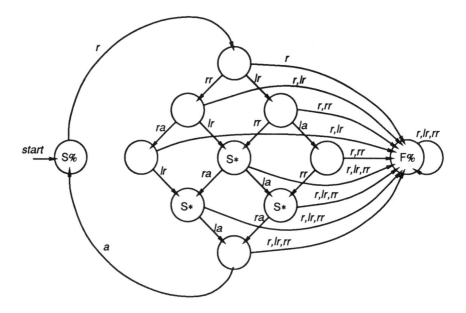

**Figure 7.6.**   Live Fork/Join

Given this example, it is not totally surprising that complete trace structures can be used to require a concurrent implementation of the fork/join. If the same state graph is used as for the prefix-closed specification, but the accepting states are chosen to represent the obvious liveness properties of the fork/join (that it eventually produces certain outputs in response to certain inputs), the automata of Figures 7.6 and 7.7 are obtained for the fork/join and sequence. Using these trace structures, the sequence *does not* conform to the fork/join: the trace *r; lr* is *possible* in the sequence, but *impossible* in the fork/join.

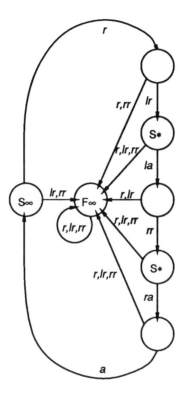

**Figure 7.7.**  Live Sequence

This accounts for the difference in deadlock behavior in the example of the previous paragraph.

Intuitively, it would seem that there are some contexts in which a sequence *could* legitimately be substituted for a concurrent fork/join. For example, if each of the request/acknowledge pairs for the left and right halves of the environment were required to be live even in the absence of a request on the other pair, it would be impossible for the sequence module to deadlock. The trace structure of Figure 7.7 can be modified to include this assumption, as in Figure 7.8 (in which the newly forbidden environmental

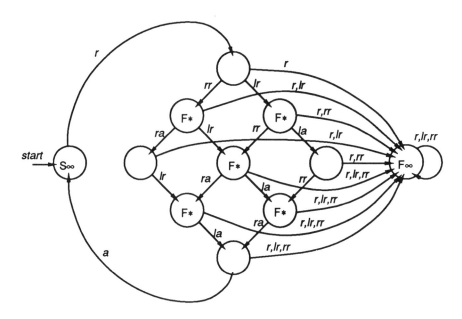

**Figure 7.8.**   Fork/Join with Live Environment

traces have been made into failures). The trace structure for the sequence module is similarly modified in Figure 7.9.

It is easy to see that the modified sequence module conforms to the modified fork/join, since the sets of failures and possible traces of the first are included in the sets of failures and possible traces of the second.

## 7.6.   Complete Trace Structures as a Lattice

Interestingly, complete trace structures as defined are *not* a lattice. The problem is that, given complete trace structures $T$ and $T'$, there is not always a complete trace structure that conforms to both. As a simple example of this, consider $T = (\emptyset, \{a, b\}, [a^\omega], \emptyset)$ and $T' = (\emptyset, \{a, b\}, [b^\omega], \emptyset)$. The sets of possible traces ($a^\omega$ and $b^\omega$) are disjoint, so there is no way that a canonical trace structure with a non-empty $P$ set can conform to both. For complete trace structures, the trace structure $(I, O, I^\infty, \emptyset)$ is *not* a universal implementation, because it may fail to meet liveness requirements.

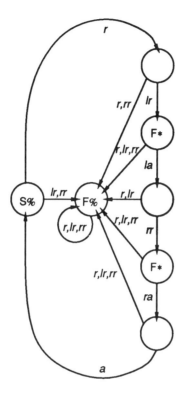

**Figure 7.9.**   Sequence with Live Environment

The meet operation was well-defined for prefix-closed trace structures because a circuit which was allowed to output neither $a$ nor $b$ would still have $\epsilon$ as a possible trace. With the added expressive power of liveness conditions, it is possible to add constraints until there is no trace structure that satisfies them.

In the case of the join operation, there *is* a universal implementation for any input and output sets: $(I, O, A^\infty, A^\infty)$ (we call this $\top$ for convenience in on of the proofs below). However, the identity of Section 4.7, that $T \sqcup T'$ has $P \cup P'$ as its set of possibilities and $F \sqcup F'$ as its failures, fails to hold for complete trace structures. The reason for this is that $T \sqcup T'$ is defined to be a *canonical* trace structure, but taking the unions of the possibilities and failures may yield a non-canonical trace structure by introducing autofailures.

For each $I$ and $O$ set, it would be possible to *add* an artificial universal implementation, written $\bot$, to the set of complete trace structures with those inputs and outputs. With this addition, the canonical complete trace structures are a lattice. The join can be computed by taking the unions of the possibility and failure sets (as before), *then reducing to canonical form*. To compute the meet, we first define a variant of the mirror operation which is defined even for the universal specification (which has $S = \emptyset$). For any canonical complete trace structure, $T$, let $\bot^X = \top$ $\top^X = \bot$ $T^X = T^M$ when $T$ is neither $\top$ nor $\bot$. With these definitions, the identity $T \sqcap T' = (T^X \cap T'^X)^X$ holds, so the meet can be computed by using the join and mirroring.

**Theorem 7.3.** *The set of canonical complete trace structures having inputs $I$ and outputs $O$ and augmented with $\bot$ is a bounded lattice.*

This theorem is proved through the following lemmas, which also show that an effective procedure exists for computing the meet and join.

**Lemma 7.12.** *Let $T$ and $T'$ be complete trace structures such that $A = A'$ and $T_S$ and $T'_S$ are the canonical complete trace structures that result from applying the simplifications to $T$ and $T'$. Then $P \subseteq P'$ and $F \subseteq F'$ implies $P_S \subseteq P'_S$ and $F_S \subseteq F'_S$.*

**proof.** By 4.6, $T \preceq T'$. The transformations preserve conformation, so $T_S \preceq T'_S$. But $T_S$ and $T'_S$ are in canonical form, so by 4.23, $P_S \subseteq P'_S$ and $F_S \subseteq F'_S$.  $\square$

Another way to look at Lemma 7.12 is that the simplifications find the trace structure with the largest $P$ and $F$ sets that is conformation-equivalent to the original. Note that Lemma 7.12 also applies to prefix-closed trace structures.

**Lemma 7.13.** *If $T$ and $T'$ are any complete trace structures, and if $T'' = (I, O, (P \cup P') - (F \cup F'), F \cup F')$, then $T \preceq T''$ and $T' \preceq T''$.*

**proof.** Let $T'' = (I, O, (P \cup P') - (F \cup F'), F \cup F')$. We claim that $T''$ is a (not necessarily canonical) complete trace structure. Everything is obvious except receptiveness. Let $x$ be any member of **pref**$(P'')$. Then, by property 6.2, **pref**$(P'') = $ **pref**$(P) \cup $ **pref**$(P')$, so $x \in $ **pref**$(P)$ or $x \in $ **pref**$(P')$. Assume without loss of generality that $x \in $ **pref**$(P)$. Since $T$ is receptive, the game $(I^*, O \cup \{\epsilon\}, \textbf{flat}^{-1}[\textbf{suf}(x, P)])$ is winnable by black. But then by property 6.6, **suf**$(x, P) \subseteq $ **suf**$(x, P'')$, so by the properties of inverse images and property G1, the game $(I^*, O \cup \{\epsilon\}, \textbf{flat}^{-1}[\textbf{suf}(x, P'')])$ is black-winnable, also. Hence, $T''$ is receptive, so it is a complete trace structure.  $\square$

**Lemma 7.14.** *Let $T, T'$, and $T'''$ be any canonical complete trace structures such that $T \preceq T'''$ and $T' \preceq T'''$, and let $T'' = (I, O, (P \cup P') - (F \cup F'), F \cup F')$. Then $T'' \preceq T'''$.*

**proof.** Since $T, T'$, and $T'''$ are canonical, $P \subseteq P'''$ and $P' \subseteq P'''$, by 4.23, so $P \cup P' \subseteq P'''$ and $F \cup F' \subseteq F'''$. Hence, by 4.6, $T'' \preceq T'''$.  $\square$

**Lemma 7.15.** *The result of simplifying* $(I, O, (P \cup P') - (F \cup F'), F \cup F')$ *to canonical form is* $T \sqcup T'$

**proof.** The simplifications preserve conformation equivalence, so the canonical form will satisfy the definition of $T \sqcup T'$, by the previous two lemmas.    □

**Lemma 7.16.** *If* $T$ *and* $T'$ *are any canonical complete trace structures and* $T \preceq T'$, *then* $T'^X \preceq T^X$.

**proof.** In case $T$ or $T'$ is the universal specification or the universal implementation, the result is immediate. Otherwise, $T^X = T^M$ and $T'^X = T'^M$. Hence, $P^M = A^\infty - F$ and $P'^M = A^\infty - F'$ (by definition), and $F \subseteq F'$ (by Lemma 4.23), so $P'^M \subseteq P^M$. Similarly, $F'^M \subseteq F^M$. Therefore, by Lemma 4.6, $T'^M \preceq T^M$.    □

**Lemma 7.17.** *For any complete canonical trace structures* $T$ *and* $T'$ *such that* $A = A'$, $T \sqcap T' = (T^X \sqcup T'^X)^X$.

**proof.** The $X$ operation reverses the sense of the conformation relations in the properties of join, yielding the properties of meet, directly.    □

The meet and join can be effectively computed because the all of the operations involved (union, intersection, and complementation of mixed-regular sets and simplification to canonical form) are effective.

## 7.7.   Practical Considerations

It would be nice to have a program just like the one of Chapter 5 for complete traces. Such a program has not been implemented. This section discusses a possible implementation approach and points out some of the problems. Most of the computations involved are analogous to the those for prefix-closed trace structures, but slower (but not exponentially slower) and more difficult to implement because the operations and decision procedures for finite automata on infinite sequences are more complicated. However, the decision procedure for infinite games, which is used in checking receptiveness and in the simplifications, is substantially worse (by several exponentials) than the corresponding computations for prefix-closed trace structures.

Mixed regular trace sets can be represented using finite automata and nondeterministic Büchi automata. There are a variety of other automata that could be used, but nondeterministic Büchi automata are relatively succinct, and operations on them are within a polynomial factor of being as fast as possible (in the absence of a subexponential solution to PSPACE-hard problems). In comparing the computations for prefix-closed and complete trace structures, it is not unreasonable for the latter to be exponentially more expensive (in the worst case) because deterministic automata were used in the implementation of prefix-closed trace structures. Even for finite sequences, deterministic automata are exponentially larger in the worst case than nondeterministic automata.

It is not difficult to implement the operations **compose, hide,** and **rename** for complete trace structures. They require doing deletion, inverse deletion, and intersection on finite and Büchi automata, which are all polynomial-time operations. Conformation can be tested by computing the mirror of the specification, composing it with the implementation, and testing for emptiness. To compute the mirror, the $P$ set needs to be complemented. The best known algorithm for complementing Büchi automata, is exponential [67] (the problem is PSPACE-hard). It may seem that the need to complement the $P$ set of the specification makes this procedure exponentially more costly in the size of the specification than the procedure for prefix-closed trace structures. However, the reachability construction in Chapter 5 is simple only because the specification is represented by a *deterministic* automaton. If a nondeterministic automaton were used, it would have to be converted to a deterministic automaton or complemented in some other way, incurring an exponential time cost in the worst case.

The sticking point is the decision procedure for black-winnability of infinite games. One problem is that it is difficult to check that a complete trace structure is well-formed; as the definitions are formulated, testing for receptiveness requires deciding whether a set of infinite games are winnable by black. Also, the simplification to canonical form is very expensive since autofailure manifestation requires testing black-winnability of infinite games, also.

The decision procedure on infinite games can perhaps be avoided by limiting the capabilities of the program. Instead of checking for receptiveness, the system could trust the user to supply well-formed trace structures. The user would have to show that the trace structures are well-formed by other means (perhaps manually). This was not difficult to do for the examples of Section 7.5. Or perhaps a restricted class of complete trace structures that is easier to check could be identified and used. For straightforward verification, in which a circuit expression is compared with a specification, autofailure manifestation is not required. However, implementing the features of Section 5.6 requires being able to do autofailure manifestation, so this would be an unfortunate limitation.

Counting exponentials is a very crude measure of performance. Some of the algorithms on Büchi automata are less efficient than the corresponding procedures on finite automata by polynomial factors or large constants. There are additional questions about expected behavior of the algorithms on "typical" circuits. For example, the program of Chapter 5 was often able to discover bugs after examining a small fraction of the global state graph for a circuit. Whether this is possible for complete traces, even disregarding the problems with infinite games, remains to be seen.

In summary, the major stumbling block to implementing this chapter is the decision procedure for infinite games. For a practical implementation, either the capabilities of the program will have to be limited, or a better decision procedure for black-winnability of infinite games will have to be found.

# Chapter 8

# Conclusion

## 8.1. Summary

We have built, from the ground up, a theory and practice for automatic verification of speed-independent circuits. The first task was the definition of a small set of operations for making circuits out of less complicated circuits, and an exploration of the identities that should hold for the operations. This gave us the concept of a *circuit algebra*. The next step was a formal model for the behavior of speed-independent circuits, based on prefix-closed trace structures. We then derived from the idea of safe substitution a way of *specifying* circuits using trace structures (the same formalism that was used for modeling). The theory was put to the test in the form of a program that could verify (and find bugs in) circuits published by other authors. Finally, the trace theory was extended to handle liveness properties in addition to safety properties, by considering non-prefix-closed sets of finite and *infinite* traces.

Another way to survey the ground covered by this thesis is to consider the different interpretations of circuit algebra and their associated equivalence relations. For example, the first interpretation was the algebra of circuit structures; the operations were applied to graph representations of circuit topologies. The equivalence relation was *structural equivalence*, which is a type of graph isomorphism. There were two behavioral interpretations: *prefix-closed trace structures*, and *complete trace structures*. These interpretations were used to model circuit operation. In each case, the relevant equivalence relation was *behavioral equivalence* — two circuits were considered equivalent if their trace sets were equal. Behavioral equivalence for complete trace structures is a finer equivalence relation than behavioral equivalence for prefix-closed trace structures. The two remaining interpretations were the algebras of *canonical prefix-closed trace structures* and *canonical complete trace structures*. In these cases, the relevant equivalence was *conformation equivalence.* Two circuits are conformation-equivalent if they are equally useful for correct circuits. Within the class of prefix-closed or complete trace structures, behavioral equivalence is a finer equivalence relation than conformation equivalence, since (for example) conformation equivalence does not distinguish between a circuit that *must fail* for a particular input and a circuit that *might fail*. Also, conformation equivalence is a finer equivalence relation for complete trace structures than for prefix-closed trace structure.

## 8.2.  Future Work

As with all research, this thesis raises more questions than it answers. This section surveys a few areas for further study.

### 8.2.1. Efficiency

First, there are a number of of improvements that should be made to the program in order to make it a more useful and efficient tool. An area which obviously needs to be investigated is that of description languages for specifying trace sets, and regular languages in general. Fortunately, it should be straightforward to connect different syntactic front-ends onto the program. Perhaps the best choice would be to support a variety of description languages, in the likely event that no language is universally superior for all applications.

Many more examples of speed-independent circuits should be tried, in order characterize the typical characteristics of circuits. It would be interesting and useful to characterize the differences between circuits that are easy to verify and circuits that are difficult.

It is likely that some circuits are difficult because their behavior is inherently complex. However, there are circuits that will cause the program of Chapter 5 to explore exponentially many states, even though the actual behavior of the circuit seems to be quite simple. For example, a composition of $n$ completely independent ring oscillators requires a state graph with $2^n$ states. The problem is that the program explicitly represents and stores all global states. A promising alternative is *symbolic* state representations, in which a small data structure (such as a logical expression) can represent a large number of states. It is not hard to invent symbolic representations that compactly represent compositions of independent circuits. However, it is much less clear what would constitute a good representation for systems that have even a small amount interaction among processes.

It is important to be able to handle liveness properties. The theory of Chapter 7 is a good start on this, but there are problems that must be solved before it can be implemented. First among these is the decision procedure for infinite games, which is impractical. If this problem can be solved or circumvented, it is likely, but not certain, that a practical verifier can be implemented. While the algorithms and decision procedures for automata on infinite sequences are not exponentially worse than the corresponding procedures for automata on finite strings, they are often significantly more expensive: for example, complementation of a nondeterministic finite automaton on finite sequences is an $O(2^n)$ operation, as compared with the $O(16^{n^2})$ operation for nondeterministic Büchi automata. When $n = 10$, this is the difference between $2^{10}$ and $2^{400}$. Also, a major feature of the approach used in Chapter 5 is that it can check for failures as it constructs the states; this is easy to do this during depth-first search. Finding failures in a finite automaton on infinite sequences requires searching for cycles of certain types in the graph. Consequently, the algorithm for automata on infinite sequences has a much more complicated control structure, which may make it more difficult to find failures without constructing many states.

### 8.2.2. Extending the Class of Circuits

Not all asynchronous circuits are speed-independent. Many actual designs depend on real-valued timing constraints of various kinds. For example, Seitz [69] and Brzozowski and Yoeli [19] propose that circuits be designed with an assumption that the ratio of the speeds of the fastest and slowest components in the circuit is not greater than $m/n$ for some integers $m > n$ (such as $3/2$ or $2/1$). Also, it may be reasonable to assume that the delays in components have constant upper and lower bounds. Such assumptions restrict the traces that occur in a composition, so that the circuit appears to satisfy more properties than it would under a speed-independent model. It would be very beneficial in practice to be able to take such assumptions into account in verifying a circuit.

Another useful extension would be the ability to handle circuits designed at levels below the gate and element level. In particular, many VLSI designers work directly at the *switch* level, exploiting the bilateral properties of MOS transistors to achieve much more efficient designs. It would also be important in practice to be able to verify circuits designed at the switch-level.

### 8.2.3. Application to Other Types of Concurrent Systems

Modularity in verifying other types of concurrent systems is just as important as for speed-independent circuits. For example, the theory developed in this thesis should apply almost directly to systems that communicate via shared distributed variables (i.e., no variable can be written by more than one process); this model is very similar to our model of wire communication in asynchronous circuits. Also, the idea of safe substitution as a basis for hierarchical verification would seem to apply to many other types of models of concurrency, including tree-based models such as CCS.

# References

[1] S. Aggarwal, R.P. Kurshan, K. Sabnani. A Calculus for Protocol Specification and Validation. In H. Rudin and C.H. West (eds), *Protocol Specification, Testing, and Verification, III*. pp. 19–34, Elsevier Science Publishers B.V. (North-Holland), 1983.

[2] S. Aggarwal, R.P. Kurshan, D. Sharma. A Language for the Specification and Analysis of Protocols. In H. Rudin and C.H. West (eds), *Protocol Specification, Testing, and Verification, III*. pp. 35–50, Elsevier Science Publishers B.V. (North-Holland), 1983.

[3] Geoff Barrett. *Formal Methods Applied to a Floating Point Number System*. Technical Monograph PRG-58, Oxford University Computing Laboratory, Programming Research Group, January, 1987.

[4] Howard Barringer, Ruurd Kuiper, and Amir Pnueli. Now You May Compose Temporal Logic Specifications. In *Proceedings of the 16th ACM Symposium on Theory of Computing*. pp. 51–63, Association for Computing Machinery, 1984.

[5] Harry G. Barrow. VERIFY: A Program for Proving Correctness of Digital Hardware Designs. *Artificial Intelligence* 24, 1984.

[6] Mark Joseph Bennett. *Proving Correctness of Asynchronous Circuits Using Temporal Logic*. PhD Thesis, University of California, Los Angeles, 1986.

[7] Gerard Berthelot and Richard Terrat. Petri Nets Theory for the Correctness of Protocols. In Carl Sunshine (ed), *Protocol Specification, Testing, and Verification, II*. pp. 325–342, Elsevier Science Publishers B.V. (North-Holland), 1982.

[8] J. Billington and M.C. Wilbur-Ham. Automated Protocol Verification. In M. Diaz (ed), *Protocol Specification, Testing, and Verification, II*. pp. 77–100, Elsevier Science Publishers B.V. (North-Holland), 1982.

[9] David L. Black. On the existence of delay-insensitive fair arbiters: Trace theory and its limitations. *Distributed Computing* 1(4):205–225, 1986.

[10] Gregor V. Bochmann. Finite State Description of Communication Protocols. *Computer Networks* 2(4,5):361–372, October, 1978.

[11] Gregor V. Bochmann. Hardware Specification with Temporal Logic: An Example. *IEEE Transactions on Computers* C-31(3):223–231, March, 1982.

[12] A. Bourguet. A Petri Net Tool for Service Validation in Protocol. In B. Sarikaya (ed), *Protocol Specification, Testing, and Verification, VI.* pp. 231–292, Elsevier Science Publishers B.V. (North-Holland), 1987.

[13] Ed Brinksma. A Tutorial on LOTOS. In M. Diaz (ed), *Protocol Specification, Testing, and Verification, V.* pp. 171–194, Elsevier Science Publishers B.V. (North-Holland), 1986.

[14] Stephen D. Brookes. *A Model for Communicating Sequential Processes.* Technical Report CMU-CS-83-149, Department of Computer Science, Carnegie-Mellon University, 1983.

[15] Michael C. Brown, Edmund M. Clarke, David L. Dill, and Bud Mishra. Automatic Verification of Sequential Circuits Using Temporal Logic. *IEEE Transactions on Computers* C-35(12):1035–1044, December, 1986.

[16] Michael C. Browne, Edmund M. Clarke, and David L. Dill. Automatic Circuit Verification Using Temporal Logic: Two New Examples. In George J. Milne and P.A. Subrahmanyam (eds), *Formal Aspects of VLSI Design, Proceedings of the 1985 Edinburgh Workshop on VLSI.* pp. 113–124, North-Holland, 1986.

[17] Erik Brunvand. PhD Thesis Proposal. July, 1987.

[18] Randal E. Bryant. Symbolic Verification of MOS Circuits. In Henry Fuchs (ed), *1985 Chapel Hill Conference on Very Large Scale Integration.* pp. 419–438, Computer Science Press, Inc., 1985.

[19] Janusz A. Brzozowski and Micheal Yoeli. *Digital Networks.* Prentice-Hall, Inc., 1976.

[20] J. Richard Büchi. On a Decision Method in Restricted Second Order Arithmetic. In *Proceedings of the 1960 International Congress on Logic, Methodology, and Philosophy of Science.* pp. 1–11, Stanford University Press, 1962.

[21] J. Richard Büchi and Lawrence H. Landweber. Solving Sequential Conditions by Finite-State Strategies. *Transactions of the American Mathematical Society* 138:295–311, April, 1969.

[22] Luca Cardelli. *An Algebraic Approach to Hardware Description and Verification.* PhD Thesis, University of Edinburgh, 1982.

[23] T.J. Chaney and C.E. Molnar. Anomalous Behavior of Synchronizer and Arbiter Circuits. *IEEE Transactions on Computers* C-22(4):421–422, April, 1973.

[24] Yaacov Choueka. Theories of Automata on $\omega$-Tapes: A Simplified Approach. *Journal of Computer and System Sciences* 8(2):117–141, April, 1974.

[25] Tam-Anh Chu. On the Models for Designing VLSI Asyncronous Digital Systems. *INTEGRATION, the VLSI journal*(4).99–113, 1986.

[26] Alonzo Church. Logic, Arithmetic, and Automata. In *Proceedings of the International Congress of Mathematicians, 1962.* pp. 23–35, Institut Mittag-Leffler, 1963.

[27] E.M. Clarke, E.A. Emerson, and A.P. Sistla. Automatic Verification of Finite-State Concurrent Systems Using Temporal Logic. *ACM Transactions on Programming Languages and Systems* 8(2):244–263, April, 1986.

[28] Wesley A. Clark and Charles E. Molnar. *Macromodular System Design.* Technical Report 23, Computer Systems Laboratory, Washington University, April, 1973.

[29] D.L. Dill and E.M. Clarke. Automatic Verification of Asynchronous Circuits Using Temporal Logic. *IEE Proceedings, Pt. E* 133(5):276–282, September, 1986.

[30] Jo C. Ebergen. Private Communication.. 1984.

[31] J.C. Ebergen. *A Technique to Design Delay-Insensitive VLSI Circuits.* Report CS–R8622, Centrum voor Wiskunde en Informatica, June, 1986.

[32] Samuel Eilenberg. *Automata, Languages, and Machines, Vol. A.* Academic Press, 1974.

[33] William I. Fletcher. *An Engineering Approach to Digital Design.* Prentice-Hall, Inc., 1980.

[34] Masahiro Fujita, Hedehiko Tanaka, Tohru Moto-oka. Verification with Prolog and Temporal Logic. In T. Uehara and M. Barbacci (eds), *IFIP Sixth Computer Hardware Description Languages and their Applications.* pp. 103–114, North-Holland Publishing Company, 1983.

[35] M.J.C. Gordon and J. Herbert. Formal Hardware Verification Methodology and its Application to a Network Interface Chip. *IEE Proceedings, Pt. E* 133(5):255–270, September, 1986.

[36] Alan B. Hayes. Stored State Asynchronous Sequential Circuits. *IEEE Transactions on Computers* C–30(8):596–600, August, 1981.

[37] Frederick J. Hill and Gerald R. Peterson. *Introduction to Switching Theory and Logical Design.* John Wiley and Sons, 1981.

[38] C.A.R. Hoare. Communicating Sequential Processes. *Communications of the Association for Computing Machinery* 21(8):666–677, August, 1978.

[39] C.A.R. Hoare. *A Model for Communicating Sequential Processes.* PRG-22, Programming Research Group, Oxford University Computing Laboratory, 1981.

[40] C.A.R. Hoare. *Communicating Sequential Processes.* Prentice-Hall, Inc., 1985.

[41] C.A.R. Hoare, S.D. Brookes, and A.W. Roscoe. *A Theory of Communicating Sequential Processes.* Technical Monograph PRG–16, Oxford University Computing Laboratory, Programming Research Group, May, 1981.

[42] Gerard J. Holzmann. Automated Protocol Validation in Argos, Assertion Proving and Scatter Searching. In Yechiam Yemini (ed), *Current Advances in Distributed Computing and Communications.* pp. 163–188, Computer Science Press, Inc., 1987.

[43] Gerard J. Holzmann. Tracing Protocols. In Yechiam Yemini (ed), *Current Advances in Distributed Computing and Communications.* pp. 189–207, Computer Science Press, Inc., 1987.

[44] John E. Hopcroft and Jeffrey D. Ullman. *Introduction to Automata Theory, Languages, and Computation.* Addison-Wesley Publising Company, 1979.

[45] David A. Huffman. The Design and Use of Hazard-Free Switching Networks. *Journal of the Association for Computing Machinery* 4(41):47–62, January, 1957.

[46] Nancy A. Lynch and Mark R. Tuttle. *Hierarchical Correctness Proofs for Distributed Algorithms.* MIT/LCS/TR-387, MIT Laboratory for Computer Science, April, 1987.

[47] Yonatan Malachi and Susan S. Owicki. Temporal Specifications of Self-Timed Systems. In H.T. Kung, Bob Sproull, and Guy Steele (eds), *CMU Conference on VLSI Systems and Computations.* pp. 203–212, Computer Science Press, Inc., 1981.

[48] Z. Manna and A. Pnueli. Temporal Verification of Concurrent Programs: The Temporal Framework for Concurrent Programs. In Robert S. Boyer and J. Strother Moore (eds), *The Correctness Problem in Computer Science.* pp. 215–273, Academic Press, 1981.

[49] Alain J. Martin. The Design of a Self-timed Circuit for Distributed Mutual Exclusion. In Henry Fuchs (ed), *1985 Chapel Hill Conference on Very Large Scale Integration.* pp. 245–260, Computer Science Press, Inc., 1985.

[50] Alain J. Martin. *A Delay-Insesitive Fair Arbiter.* Technical Report 5193:TR:85, Computer Science Department, California Institute of Technology, 1985.

[51] Carver Mead and Lynn Conway. *Introduction to VLSI Systems.* Addison-Wesley Publishing Company, 1980.

[52] Teresa H.-Y. Meng and Robert W. Brodersen and David G. Messerschmitt. Automatic synthesis of Asynchronous Circuits from High-Level Specifications. September.

[53] Raymond E. Miller. *Switching Theory. Volume II: Sequential Circuits and Machines.* John Wiley and Sons, 1965.

[54] Robin Milner. *Lecture Notes in Computer Science. Volume 92: A Calculus of Communicating Systems.* Springer-Verlag, 1980.

[55] B. Mishra and E.M. Clarke. Hierarchical Verification of Asynchronous Circuits Using Temporal Logic. *Theoretical Computer Science* 38:269–291, 1985.

[56] Jayadev Misra and K. Mani Chandy. Proofs of Networks of Processes. *IEEE Transactions on Software Engineering* SE–7(4), July, 1981.

[57] Charles E. Molnar, Ting-Pien Fang, and Frederick U. Rosenberger. Synthesis of Delay-Insensitive Modules. In Henry Fuchs (ed), *1985 Chapel Hill Conference on Very Large Scale Integration*. pp. 67–86, Computer Science Press, Inc., 1985.

[58] David E. Muller and W.S. Bartky. A Theory of Asynchronous Circuits. In *The Annals of the Computation Laboratory of Harvard University. Volume XXIX: Proceedings of an International Symposium on the Theory of Switching, Part I*. pp. 204–243, Harvard University Press, 1959.

[59] Suhas S. Patil. *An Asynchronous Logic Array*. Technical Memorandom 62, Massachusetts Institute of Technology, Project MAC, 1975.

[60] James Lyle Peterson. *Petri Net Theory and the Modeling of Systems*. Prentice-Hall, Inc., 1981.

[61] A. Pnueli. The Temporal Logic of Programs. In *18th IEEE Symposium on Foundations of Computer Science*. pp. 46–57, IEEE, 1977.

[62] Amir Pnueli. In Transition from Global to Modular Temporal Reasoning about Programs. In Kzysztof Apt (ed), *NATO ASI Series F: Computer and System Sciences. Volume 13: Logics and Models of Concurrent Systems*. pp. 123–144, Springer-Verlag, 1985.

[63] Amir Pnueli and Roni Rosner. On the Synthesis of a Reactive Module. In *16th ACM Symposium on Principles of Programming Languages*. pp. 179–190, ACM, 1989.

[64] Michael O. Rabin. Weakly Definable Relations and Special Automata. In Yehoshua Bar-Hillel (ed), *Mathematical Logic and Foundations of Set Theory*. pp. 1–23, North-Holland Publishing Company, 1970.

[65] Michael O. Rabin. *Regional Conference Series in Mathematics. Volume 13: Automata on Infinite Objects and Church's Problem*. American Mathematical Society, 1972.

[66] Martin Rem, Jan L.A. van de Snepscheut, and Jan Tijmen Udding. Trace Theory and the Definition of Hierarchical Components. In Randal Bryant (ed), *Third Cal-Tech Conference on Very Large Scale Integration*. pp. 225–239, Computer Science Press, Inc., 1983.

[67] Shmuel Safra. On the Complexity of $\omega$-automata. In *29th IEEE Symposium on Foundations of Computer Science*. pp. 319–327, IEEE, 1988.

[68] Charles L. Seitz. Asynchronous Machines Exhibiting Concurrency. In *Project MAC Conference on Concurrent Systems and Parallel Computation*. pp. 93–106, Association for Computing Machinery, June, 1970.

[69] Charles L. Seitz. Chapter 7: System Timing. In C. Mead and L. Conway (eds), *Introduction to VLSI Systems*. pp. 218–262, Addison-Wesley, 1980.

[70] Charles L. Seitz. Ideas About Arbiters. *Lambda*:10–14, First Quarter, 1980.

[71] A.P. Sistla, M.Y. Vardi, and P. Wolper. The Complementation Problem for Büchi Automata with Applications to Temporal Logic. In W. Brauer (ed), *Lecture Notes in Computer Science. Volume 194: Automata, Languages, and Programming*. pp. 465–474, Springer-Verlag, 1985.

[72] Jan L.A. van de Snepscheut. Deriving Circuits from Programs. In Randal Bryant (ed), *Third CalTech Conference on Very Large Scale Integration*. pp. 241–256, Computer Science Press, Inc., 1983.

[73] Jan L.A. van de Snepscheut. *Trace Theory and VLSI Design*. PhD Thesis, Department of Computing Science, Eindhoven University of Technology, October, 1983.

[74] Guy L. Steele, Jr. *Common LISP: The Language*. Digital Press, 1984.

[75] B.A. Trakhtenbrot and Y.A. Barzdin. *Fundamental Studies in Computer Science. Volume 1: Finite Automata: Behavior and Synthesis*. North-Holland Publishing Company, 1973.

[76] Jan Tijmen Udding. A formal model for defining and classifying delay-insensitive circuits and systems. *Distributed Computing* 1(4):197–204, 1986.

[77] Jan Tijmen Udding. *Classification and Composition of Delay-Insensitive Circuits*. PhD Thesis, Department of Computing Science, Eindhoven University of Technology, September, 1984.

[78] Stephen H. Unger. *Asynchronous Sequential Switching Circuits*. Robert E. Kreiger Publishing Company, Inc., 1969.

[79] Kaes van Birkel. Lecture at the Workshop on Design and Implementation of Concurrent Programs, Groningen, The Netherlands, 17–21 November, 1986.

[80] Kevin S. Van Horn. *An Approach to Concurrent Semantics Using Complete Traces*. Technical Report 5236:TR:86, Computer Science Department, California Institute of Technology, 1986.

[81] Moshe Y. Vardi. Automatic Verification of Probabilistic Concurrent Finite-State Programs. In *26th IEEE Symposium on Foundations of Computer Science*. pp. 327–338, IEEE, 1985.

[82] David Winkel and Franklin Prosser. *The Art of Digital Design*. Prentice-Hall, Inc., 1980.

[83] Pierre Wolper. Temporal Logic Can be More Expressive. In *22nd IEEE Symposium on Foundations of Computer Science.* pp. 340–348, IEEE, 1981.

[84] Pitro Zafiropulo, Colin H. West, Harry Rudin, D.D. Cowan, Daniel Brand. Towards Analyzing and Synthesizing Protocols. *IEEE Transactions on Communications* COM-28(4), April, 1980.

[85] Job Zwiers, Willem Paul De Roever, and Peter van Emde Boas. Compositionality and Concurrent Networks: Soundness and Completeness of a Proofsystem. In W. Brauer (ed), *Lecture Notes in Computer Science. Volume 194: Automata, Languages, and Programming.* pp. 509–519, Springer-Verlag, 1985.

# Index

*The MIT Press, with Peter Denning, general consulting editor, and Brian Randall, European consulting editor, publishes computer science books in the following series:*

**ACM Doctoral Dissertation Award and Distinguished Dissertation Series**

**Artificial Intelligence**, Patrick Henry Winston and J. Michael Brady founding editors; J. Michael Brady, Daniel G. Bobrow, and Randall Davis, current editors

**Charles Babbage Institute Reprint Series for the History of Computing**, Martin Campbell-Kelly, editor

**Computer Systems**, Herb Schwetman, editor

**Exploring with Logo**, E. Paul Goldenberg, editor

**Foundations of Computing**, Michael Garey and Albert Meyer, editors

**History of Computing**, I. Bernard Cohen and William Aspray, editors

**Information Systems**, Michael Lesk, editor

**Logic Programming**, Ehud Shapiro, editor; Fernando Pereira, Koichi Furukawa, and D. H. D. Warren, associate editors

**The MIT Electrical Engineering and Computer Science Series**

**Research Monographs in Parallel and Distributed Processing,** Christopher Jesshope and David Klappholz, editors

**Scientific Computation**, Dennis Gannon, editor

**Technical Communication**, Edward Barrett, editor